Geostatistics for the Mining Industry

Geostatistics for the Mining Industry

Applications to Porphyry Copper Deposits

Xavier Emery

Department of Mining Engineering, University of Chile, Santiago, Chile
Advanced Mining Technology Center, University of Chile, Santiago, Chile

Serge Antoine Séguret

Centre de Géosciences, Mines ParisTech, Fontainebleau, France

CRC Press
Taylor & Francis Group
Boca Raton London New York

CRC Press is an imprint of the
Taylor & Francis Group, an **Informa** business

Géostatistique de gisements de cuivre chiliens

par: Serge Antoine Séguret et Xavier Emery

First published by Presses des Mines, Paris, 2019,
ISBN: 978-2-35671-562-3
English title:

Geostatistics for the mining industry: Applications to porphyry copper deposits

Typeset by codeMantra

Library of Congress Cataloging-in-Publication Data
Names: Séguret, Serge Antoine, author. | Emery, Xavier, author.
Title: Geostatistics for the mining industry : 35 years of research applied to copper deposits / Xavier Emery and Serge Antoine Séguret.
Other titles: Géostatistique de gisements de cuivre chiliens. English
Description: Boca Raton : CRC Press, 2020. | Translation of: Géostatistique de gisements de cuivre chiliens (Séguret and Emery), 2019. | Includes bibliographical references and index. |
Identifiers: LCCN 2020030090 (print) | LCCN 2020030091 (ebook)
Subjects: LCSH: Copper mines and mining—Chile—Statistical methods. | Mining engineering—Statistical methods. | Geology—Statistical methods. | Kriging.
Classification: LCC TN444.C5 S4413 2020 (print) | LCC TN444.C5 (ebook) | DDC 553.4/3015195—dc23
LC record available at https://lccn.loc.gov/2020030090
LC ebook record available at https://lccn.loc.gov/2020030091

ISBN: 978-0-367-50575-2 (hbk)
ISBN: 978-1-003-05046-9 (ebk)

DOI: 10.1201/9781003050469
https://doi.org/10.1201/9781003050469

To Pedro Carrasco,
the friend who left us too soon.
To Samuel Emery and Romane Séguret,
our dear children.

Contents

Acknowledgments

Closely or remotely involved, in one way or another, but always with an essential role due to the butterfly effect, the following people have contributed to the content of this work: engineers, geologists, geostatisticians, students, professors, researchers, managers, consultants, journal editors, secretaries. We decided not to submit to the yoke of the alphabetical order; these friends appear below in the order that our memory left them:

Pedro Carrasco, Juan Enrique Morales, Fernando Geister, Julio Beniscelli, Sergio Fuentes, Nicolas Cheimanoff, Pablo Carrasco, Antonio Cortés, Jean Serra, Pierre-Yves Descote, Cristian Guajardo, Francky Fouedjio, Nicolas Desassis, Hélène Beucher, Gaëlle Le Loc'h, Felipe Ibarra, Danitza Aburto, Felipe Celay, Cristian Quiñones, José Delgado, Rodrigo Riquelme, Ghislain De Marsily and his wife, Christian Ravenne, Álvaro Herrera, Arman Melkumyan, Cristian Díaz, Serge Beucher, Sebastián De La Fuente, Ricardo Vargas, Juan Soto, Carlos Montoya, Javier Cornejo, Gonzalo Nelis, Mohammad Koneshloo, Mustafa Touati, Adrian Vargas, Bruno Behn, Aldo Casali, Hans Göpfert, Eduardo Magri, Gonzalo Montes-Atenas, Nelson Morales, Brian Townley, Nadia Mery, Mohammad Maleki, Nasser Madani, Amin Hekmatnejad, Manuel Reyes, Margaret Armstrong, Alain Galli, Leandro De Oliveira, Celeste Queiroz, Tony Wain, Francisco Farías, Marcio Soares, Juan-Carlos Alvarez, Luis Candelario, Cecilia Idelfonso, Frédéric Rambert, Jacques Deraisme, François Geffroy, Claudio Martínez, Ramón Freire, Andrés Brzovic, Paulina Schachter, Karen Baraona, Violeta Ríos, Sebastián González, María Paz Sepúlveda, Iván Rojas, Eduardo Jara, Jean-Paul Chilès, Jacques Rivoirard, Jacques Laurent, Isabelle Schmitt, Nathalie Dietrich, Verónica Möller, Denis Marcotte, Roussos Dimitrakopoulos, Guillaume Caumon, Claudio Rojas, Silvia Dekorsy and the production team at Presses des Mines, Alistair Bright, Cathy Hurren, Kritheka and the production team at CodeMantra.

Françoise, wife of Georges Matheron, their daughters Laure, Isabelle, Cécile, Lydie.

Special thanks to James Pocoe, for his meticulous reading of this English version of the book.

We also thank the institutions that made possible this work and the research it covers:

- the National Copper Corporation of Chile (Codelco);
- la University of Chile, its Mining Engineering Department and Advanced Mining Technology Center;

- the National Agency for Research and Development of Chile (ANID), through grants CONICYT PIA AFB180004, CONICYT PIA Anillo ACT 1407 (contribution to Chapter 4) and CONICYT/FONDECYT/REGULAR/N°1170101 (contribution to Chapters 6 and 7);
- Mines ParisTech and its Centre of Geosciences;
- ARMINES, the association that manages the industrial activities of Mines ParisTech, among others.

A special mention for Isatis and Matlab software, commercialized by the companies Geovariances and The MathWorks, Inc., with which most of the studies in this book were made.

A bit of history

This book is the fruit of thirty-five years of accumulated experience of two researchers: Xavier Emery, a civil engineer from Paris School of Mines, currently a full professor in the Department of Mining Engineering and a principal researcher of the Advanced Mining Technology Center (AMTC) in the Faculty of Physical and Mathematical Sciences at University of Chile; and Serge Antoine Séguret, a research engineer at Mines ParisTech (Paris School of Mines) who has been working for more than thirty-five years in the prestigious Center of Geostatistics founded by Georges Matheron (1930–2000).

Georges Matheron and geostatistics, two names that are inextricably linked. This engineer from the Corps of Mining Engineers of France is the founder of the discipline that today is increasingly called 'spatial statistics', which also includes the time axis in which the variables studied are sometimes regionalized, as well as the analysis of space–time fields, frequently found in pollution studies. The story of the discipline and of its founder begins with a field experience at the Algerian Mining Research Office (BRMA), where Georges Matheron worked in his early career.

The reading of two articles by a talented South African mining engineer, Danie Krige (1919–2013), on the evaluation of mineral resources and the support effect, published in 1951 and 1952, caused a masterful spark that led to the 'theory of regionalized variables' (Matheron, 1965a, 1971). One of the greatest contributions of this theory is a spatial prediction method, used worldwide by tens of thousands of practitioners, known as 'kriging', a name coined by Georges Matheron (1962a, b, 1963a, b, 1965a) so that nobody, ever, forgets the initial South African contribution.

But the real adventure began at Paris School of Mines in 1968, in Fontainebleau (France), where, little by little, several young researchers joined Georges Matheron to form a team of a dozen passionate people, who could not cope with a considerable demand and were also aware of contributing to the history of 'applied mathematics' at the service of engineering. This team has been constantly renewed, the two authors of this book being part of the third and fourth generations. Like their predecessors, they will have dedicated their lives to their passion and did not make geostatistics a simple job; rather, it is as if they had joined a religion.

As a result of this considerable experience, this book is an invitation to travel in several dimensions:

- the mining world with its practices;
- Chile, a pioneer country in the development and application of geostatistics, which we will travel along an axis 2,000 km along the Andes from the east of the capital, Santiago, to the Atacama Desert in the north;
- the dimension of geostatistical methods;
- finally, probably the most important, the dimension of a fantastic Franco-Chilean human adventure in which more than 100 actors from both countries participated and which began in the 1960s.

At that time, to teach students in the mining engineering courses, the Department of Mining Engineering at the University of Chile hired the services of Jacques Damay, a French engineer who was working in Santiago for the mining company Disputada de Las Condes. In 1968, under the leadership of Damay, the University of Chile and Paris School of Mines established an intense collaboration around 'geostatistics', recently created by Georges Matheron. This collaboration resulted in numerous stays and visits of Chileans in France and French in Chile, in particular, in the arrival of Alain Maréchal, a researcher at the Center of Geostatistics at Paris School of Mines, as a visiting professor at the University of Chile between 1970 and 1973. It also led to the foundation of a 'Center of Geostatistics and Mineral Deposit Evaluation' at the University of Chile, and the publication of a journal, the *Geostatistics Bulletin* (*Boletín de Geoestadística*), between 1972 and 1973 (Dagbert and Maréchal, 2008). The library of the Faculty of Physical and Mathematical Sciences at University of Chile conserves documents that witness this intense activity: course notes of statistics and mineral deposit evaluation, special issue of the magazine *Minerales* from the Institute of Mining Engineers of Chile, dedicated to geostatistics (April 1969), or even the three volumes of the mythical 'Applied Geostatistics Treaty' (*Traité de Géostatistique Appliquée*) by Georges Matheron (1962–1963) – the first two ones published as memoirs of the French Geological Survey (BRGM, *Bureau de Recherches Géologiques et Minières*), leaving the third one in the form of an unpublished note from Paris School of Mines – volumes that practically cannot be found anywhere else.

During the 1970s, Chile was a laboratory for the development of new geostatistical tools, models and methods. Witnesses of that era are the concepts of selectivity, transfer or recovery functions, discrete Gaussian model or nonlinear geostatistical techniques such as disjunctive kriging, forged at Paris School of Mines to answer the problem of calculating the recoverable resources in Chilean deposits such as Chuquicamata or El Teniente (Journel and Segovia, 1974; Maréchal, 1982). We will find these tools and techniques in this book with recent applications.

At the same time, in 1971, Chile nationalized the main copper mines that were operating in its territory, and in 1976, the National Copper Corporation of Chile (Codelco) was created to manage the nationalized deposits. There was born the largest copper producer in the world.

Paris School of Mines, the University of Chile, Codelco: three actors who would meet in the late 1990s and early 2000s, at the initiative of Chilean geologist Pedro Carrasco, the friend to whom we dedicate this book.

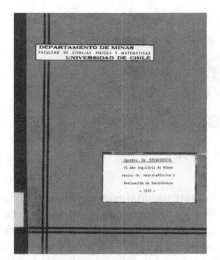

Figure 0.1 Course notes from the 'Center of Geostatistics and Mineral Deposit Evaluation' at University of Chile (1970) and the first issue of the 'Geostatistics Bulletin' (1972).

In 1979, Pedro graduated as a geologist from the University of Chile and quickly became interested in geostatistics, participating in the first session of what would be the Cycle of Specialized Training in Geostatistics (CFSG, for its acronym in French) at Paris School of Mines, a ten-month training that continues to be offered today and whose objective is to teach geostatistics to professionals with a mining experience: geologists, mining engineers, metallurgists, chemical engineers, geotechnicians and geomechanicians. Pedro then worked as a geostatistical geologist in Australia and Chile, where he became the Director of Geosciences at Codelco.

Pedro was a participant of key collaboration agreements, on the one hand, between Codelco and the University of Chile and, on the other hand, between Codelco and Paris School of Mines. Thus, a first agreement, between 1997 and 2011, allowed the establishment of an Evaluation of Mineral Deposit Chair in the Mining Engineering Department at University of Chile, including the financing of research activities and training of mining engineers and postgraduate students in mining or geology. A second agreement, between 2003 and 2016, was established with Paris School of Mines to develop solutions to mining problems, from exploration to geometallurgy, passing through sampling problems, and to implement them in the company through one-week practical workshops that gathered a dozen engineers and geologists to learn how to reproduce the proposed solutions with their data.

This book covers all these successful collaborations, of which the authors have been the main actors. Translated from French (Séguret and Emery, 2019) and Spanish (Emery and Séguret, 2020) and revised, this version brings to mining engineers and geologists our experience, through applications to porphyry copper deposits, which can certainly be extended to other types of metallic and nonmetallic deposits.

Our main motivation is to honor the memory of a friend who was the initiator of this editorial work, with such energy that made it possible to overcome his passing in

Figure 0.2 Pedro Carrasco Castelli (1949–2010).

July 2010 and whose impulse continues to motivate us, challenging the terrible laws of human entropy. We also want to thank Codelco for financing an important part of this work based on authorized publications and to convey the taste for applied research. To this end, it has been resolved to adorn the text of some anecdotes and call their actors by their names as a sign of friendship, recognition and thanks, also detailed at the end of the book.

The road has been long to get here: 35 cumulated years of applied research; several hundred thousand kilometers across Chile or the Atlantic Ocean; almost all the mining problems addressed; almost all the geostatistical techniques used, evaluated and validated; a few original innovations that advance, little by little; hundreds of friends discovered – what a wealth!

Figure 0.3 Location of the mines operated by Codelco that will be studied in this book.

Introduction

The organization of this book follows the typical mining sequence that starts with the exploration phase aimed at delineating the mineral deposit and detailing its geology.

Geological objects can be singular, like the inverted cone-shaped diatreme of the El Teniente copper mine; more generally, they can be domains with given lithological, mineralogical, structural or alteration properties, which will be designated throughout this book under the name of 'facies', commonly used in geosciences. Within each facies, the metal grades have their own distribution and variability. This peculiarity will lead directly to the prediction or simulation of the grades (quantitative variables) taking into account the statistical and spatial characteristics of each facies (categorical variable).

We emphasize these four terms, which will be found throughout this book: prediction, simulation, quantitative variable and categorical variable.

The second phase corresponds to the evaluation of the 'in situ' mineral resources. The deposit is divided into production blocks, in the order of several hundred cubic meters, for which it is necessary to predict, among other characteristics, the grades of the elements of interest, whether main products, by-products or contaminants. Predicting the next months or the next years of production involves sequencing the optimal extraction of these blocks in space and dimensioning the extraction and treatment facilities. At this point, a prediction bias can be catastrophic, especially if planned production is systematically under- or overvalued.

The 'recoverable resources' correspond to the blocks for which the grade of the main product (or an 'equivalent grade' consisting of a combination of the grades of several elements of interest) exceeds a cutoff grade that is determined taking into account numerous technical and economic conditions (which are sometimes ignored despite being mentioned in the rules and codes that regulate the public reporting of mineral resources). To correctly predict recoverable tonnage, grade and metal quantity, it is not enough to predict the average grades; one must also know their variability. Likewise, the calculations must take into account the volume of the production blocks, which is considerably larger than the exploration samples, resulting in a significant loss of selectivity, which is the ability to separate ore from waste. This problem is the subject of the third chapter.

The block model is used by the planning engineer to value the resources, to determine the final extent of the exploitation, whether open pit or underground, and to define a mining strategy that leads to a production schedule for the life of the mine. This long-term plan should not only be optimal by maximizing the net present value (NPV) but also robust against unforeseen deviations between the resources predicted by the models and the reality – the well-known geological uncertainty. At this stage, the ore reserves are defined, which correspond to the part of the in situ mineral resources that can be mined, accounting for technical and economic criteria known as 'modifying factors'.

Subsequently, in the production phase, additional sample measurements are acquired, such as blast holes in open-pit mines. These new measurements often represent larger volumes of material than those from exploration drill holes. Comparing both types of measurements by taking into account their respective volumetric supports is one of the traditional geostatistical challenges. The information support must be considered when, for example, updating the block model based on the new information from sampling in the production phase. Such additional information also allows for the definitive classification of the reserves into different types of ore and waste (a process known as 'ore control') and the short-term planning of the exploitation.

And so, the mining sequence outlines the first five chapters of this book:

1. Geometry and geology
2. Geology and grades
3. Grades and recoverable resources
4. Long-term planning and reserves
5. Short-term planning.

In addition to geology, two other disciplines, geotechnics and geometallurgy, must be addressed from the exploration stage as they play an essential role in the conception of the mine and its infrastructure and in designing the mineral concentration plant, leaching heaps or stockpiles, depending on the type of ore that will be treated.

Geotechnics concerns the behavior of the rock mass, and a major concern of all exploitation is the safety of the infrastructure and personnel working in the mine. It is necessary to study the systems of geological discontinuities, such as faults, joints or fractures, and to evaluate the quality of the rock in order to quantify as precisely as possible the degree of stability of the slopes of the exploitation in the open-pit context or the fortification of the galleries in underground mining. Geotechnics is the subject of the sixth chapter.

Geometallurgy concerns the treatment of the extracted ore. One has to assess the grade and metal quantity that will be effectively recovered after a complex process whose performance depends on numerous variables such as lithology, mineralogy, alteration, texture or grain size of the ore – variables that are often interrelated. Geometallurgy has now become a specialization in itself and is addressed in the seventh chapter.

Finally, how not to end up with sampling biases, considering here a broader meaning of the term 'sampling' given by the French engineer Pierre Gy (1924–2015) in the 1950s, in relation to the usual practice of reducing the mass of samples to a few grams whose average grade is representative of the initial lot? This sampling problem is critical to the entire mining evaluation process from the first exploration stage to obtain the information that will underpin the geological, geotechnical and geometallurgical models, through to the production stage in the ore control function involving sampling of blast holes or of underground galleries.

The three chapters are as follows, accordingly:

6. Geotechnics
7. Recovery and geometallurgy
8. Sampling and geostatistics.

The overall sequence of chapters is outlined in Figure 0.4 that reproduces the sequence of mining operations. The arrows represent the relationships or interactions between the addressed topics. At the center of the figure, the concept of geological uncertainty is indicated, essential in planning since it affects decision-making both in the long and short terms. To accompany the reader, this scheme is reproduced at the beginning of each chapter, with a highlighting of the problem addressed.

A chapter of conclusions and perspectives will allow to verify that, among all the regionalized variables under study, indicator functions play an essential role; that multivariate geostatistics and its armed wing, cokriging, despite being attractive in theory, are a complication that does not always provide a real benefit; that, regardless of the method used, the variables under study should be additive, a powerful property whose meaning goes beyond just wanting to make sense to spatial averaging. Finally, we will raise some research topics that, in our opinion, deserve to be deepened.

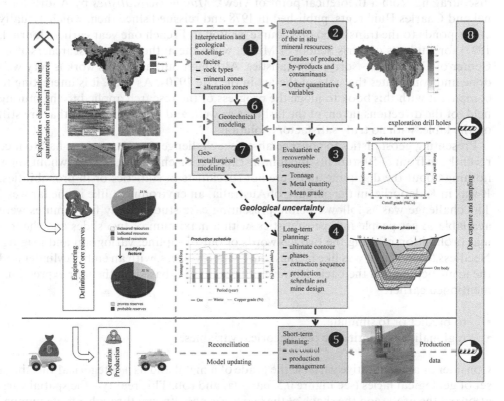

Figure 0.4 Synthetic scheme of the book plan. The boxes in the central and right columns represent the issues addressed in Chapters 1–8.

An appendix gathers some practical concepts and considerations on exploratory data analysis, structural analysis, kriging, cokriging and geostatistical simulation, as well as a presentation of two particular classes of models: transitive representations and object-based models. The novice reader is recommended to start with this appendix, but only after reading the following section.

I. 'Geostats', what is this?

This is the name used between us, much simpler, that avoids the common lapse to use the term 'geostatics' instead of the correct one, which annoys some of the most radical members of our community, especially when this erroneous term appears in a letter in which a candidate applies for a doctorate in 'geostatics', a discipline that he/she has followed with interest for several years and whose dynamism encourages him/her to... (we let you complete the sentence).

But, by the way, what is geostatistics? Although this book is mostly aimed at trained users and students, it could reach the hands of simply curious people with a good scientific level (engineers, geologists, technicians or people with a master's degree) who do not know this discipline. This section is designed for them, perhaps before orienting them toward a more complete introductory work on the subject (but not too discouraging from a theoretical point of view), *Mining Geostatistics* by André Journel and Charles Huijbregts, published in 1978 and reissued since then, which actually corresponds to the translation of a course written in French one year earlier (Journel, 1977) for students of Paris School of Mines, when the authors were researchers at its Center of Geostatics – sorry, Geostatistics. Although outdated, this work is very well done and constitutes the state-of-the-art in the late 1970s. As such, it is interesting to compare it with this book to judge the progress of the research applied to the mining field, of the directions taken, of the fashions as well, and of the problems that are still on the table, with no truly satisfactory solution.

Presenting geostatistics to nonspecialists is a challenge that the authors had to face recently, in front of 50 students and mining exploration engineers, most of whom knew nothing about the discipline and were not specialists in statistics or in probabilities. It was in Kalgoorlie, in the outback of Australia, an environment like in the far west. The challenge was as follows: before presenting case studies, only five minutes were available to get people into geostatistics, with a maximum of two slides and the formal prohibition of using the two keywords of our discipline: variogram and kriging. Needless to say that it was like facing a duel with revolvers, with an empty cylinder and the arms tied behind the back. However, the exercise is based on the four expressions mentioned earlier:

- to predict or to simulate
- regionalized quantitative or categorical variables.

Consider as a quantitative variable the grade of a metal and as a categorical variable a set of geological facies (see Figure 0.5, parts (a) and (b)). This reality – the spatial variations of the grade and the shape of the facies – is only known through a finite number of samples, which can be arranged in a regular mesh, like here, or in an irregular one, so that our knowledge is summarized by the grades and the facies of parts (c) and (d).

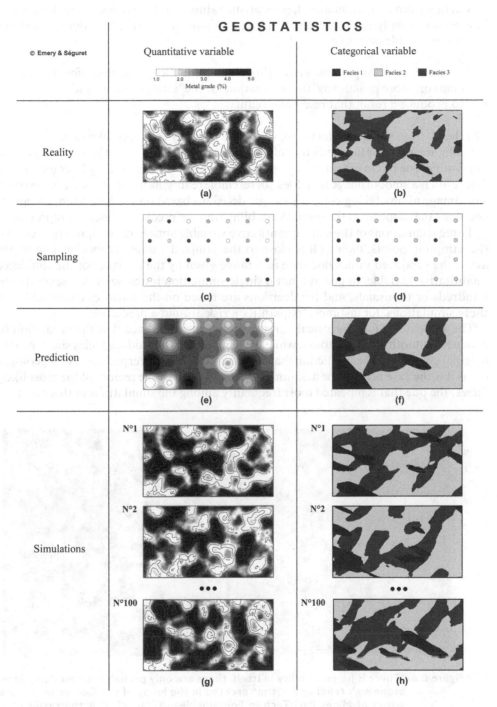

Figure 0.5 Variables distributed in a region of space or 'regionalized': metal grade (a), geological facies (b). Sampling data: metal grade (c), facies (d). A prediction of the grade (e) and the facies (f). Three simulations of the grade (g) and the facies (h).

Starting from this raw material, geostatistics aims to 'fill the holes' at any location of interest, eventually to characterize different amounts of matter, in order to meet two rarely compatible objectives:

- to make the least possible error: this objective is the 'prediction' (for the statistician) or, more prosaically, the 'evaluation' or the 'interpolation'; and
- to produce a result that resembles reality.

This leads us to subfigures (e) to (h), where the four key concepts intersect.

Subfigure (e) shows the prediction of a quantitative variable. To make an error as small as possible, one has to accept giving up the details and following the big lines (or trends). The result is a smooth image that does not resemble reality, as it does not have its 'texture'. This image allows taking certain strategic decisions based on the long term, except for decisions that depend on the spatial variability that the prediction does not reproduce.

Three simulations of the same quantitative variable appear in subfigure (g). Around the sampling points, the result is close to the sampled values; as soon as one moves away, the produced values fluctuate in a range given by the statistics of the simulated quantitative variable. In general, not a single simulation is analyzed, but several tens, hundreds or thousands, and the decisions are based on the statistics calculated over these simulations, for instance, to quantify a risk around a prediction.

The same holds for categorical variables, with the difference that the prediction (f) produces smooth facies contours, while the simulations (h) produce facies similar to the original geometry. The prediction may be the result of the interpretation of a geologist or, as it is the case here, of the assignment, at each point of the region, of the most likely facies, the one that is repeated most frequently among the simulations at this point.

Figure 0.6 'There is no probability in itself, there are only probabilistic models'. Sandstone high relief of 2×1 m^2 erected in the lobby of the Center of Geostatistics of Mines ParisTech in Fontainebleau (France). It is the result of an international subscription that brought together several hundred people and dozens of institutions in a unanimous effort to pay tribute to Georges Matheron one year after his passing.

Another term actually deserves to be added to the four previous keywords: 'modeling'. Indeed, the construction of predictions and simulations is always based on a mathematical modeling of the data. In the 'constitutive' model of geostatistics formalized by Matheron (1963b, 1965a, 1978), the regionalized variable under study is interpreted as the result (in technical language, one speaks of a 'realization') of a spatial random field, i.e., a mapping whose values are random and, at the same time, exhibit a spatial structure. To determine this abstract random field from the sampling data, several limiting hypotheses need to be introduced. Some correspond to methodological choices, and others have a more 'objective' character, in the sense that they are verifiable or falsifiable on the basis of a large number of observations. These choices and hypotheses constitute additional information to the sampling data, valuable for the prediction or the simulation of the variable under study.

Finally, let us point out that the modeling of a regionalized variable by a random field does not result in a single solution: several models are conceivable to represent the same reality with a certain degree of realism. They depend on the experience, knowledge, intuition and preferences of the geostatistician, as well as on two spatial characteristics of the variable under study:

- Its 'support', a term that refers to the elementary surface or volume on which it is measured. The support of sampling data is often assimilated to a point, although this is a simplification. A more realistic representation of a drill-hole core, for example, is a small-diameter cylinder whose length varies from a few tens of centimeters for vein-type deposits to several meters for more massive deposits and whose width varies according to drilling method. Increasing the support of the data by creating regularized samples, composited to a greater length, smoothens out the spatial variability of the regionalized variable – this is the famous 'support effect' highlighted by Danie Krige and formalized by Georges Matheron – and, therefore, modifies the probability distributions of the associated random field.
- The scale at which the regionalized variable is observed. The theory conceives the random field model in a space of two or three dimensions, without limits, but this again is an idealization because the variable is always defined in a bounded region of space, called 'field' or simply 'domain'. Depending on the objectives of the study and on the number of available samples, the geostatistician can work at the scale of the entire domain or at the scale of a local 'neighborhood' and conceive, for the same regionalized variable measured on the same volumetric support, different random field models that depend on the working scale.

Now, you know almost as much about geostatistics as we do. It is time to pack your bags and to continue this journey.

Ready? We start!

Chapter 1

Geometry and geology

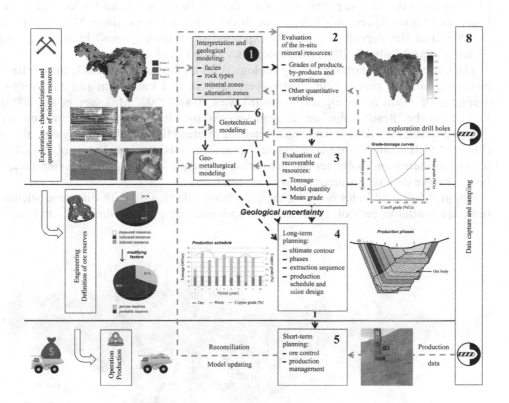

Three applications of geostatistics to geological modeling are presented. The first one refers to a simple geometry and calls into question the use or not of geostatistics. The diatreme pipe at the El Teniente mine is a chimney-like structure filled with weakly mineralized breccias, with an approximately conic shape. It can be represented by a group of measurements set to 1 (one) immersed in an infinite field of 0 (zero) that indicates its absence. In this context, is it acceptable to apply the geostatistical formalism of random fields, which always relies, in one way or another, on a form of stationarity that assumes that statistical properties such as the mean, the variance or the variogram are invariant by a translation in space?

The second and third applications focus on the modeling of mineralized breccias and of the surrounding facies, a characteristic of porphyry copper deposits. These deposits constitute the main resource of copper and molybdenum worldwide. They are often associated with porphyritic rocks that can be distinguished by the presence of feldspar phenocrystals embedded in a fine matrix. Etymologically, *porphyra* means 'purple', in reference to the best-known red variety of these rocks. Here, the challenge is to reproduce the existing natural transitions when moving from one facies to another, the way in which the rock types intermingle in space.

1.1 The 'Braden pipe' of El Teniente

El Teniente is the largest porphyry copper deposit in the world and is located in the Rancagua region of central Chile, about 70 km southeast of Santiago. Mining started in 1905, and the deposit is currently exploited from underground by *panel caving* (Hustrulid and Bullock, 2001) with more than 3,000 km of galleries.

The deposit is also famous among geologists for its diatreme, a weakly mineralized intrusion with the shape of an inverted vertical cone of 1 km height and a maximum diameter of 1 km at its highest point (Skewes et al., 2002; Maksaev et al., 2004). Known as the 'Braden pipe' or 'Braden breccia' by Chilean geologists and miners, this unit crosses the deposit and constitutes the internal limit of the mining operations (Figure 1.1). It is important to determine the surface of its contour. This problem refers to the geometry and precedes the prediction of copper grade, which is often conditioned by the defined geometry, as will be seen later.

About 4,500 drill holes crossing the pipe from side to side with different orientations are available, each of which has been split into composite samples of 6 m length.

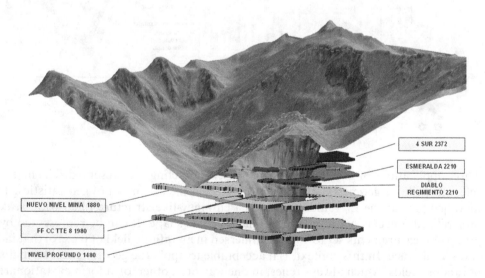

NUEVO NIVEL MINA 1880

FF CC TTE 8 1980

NIVEL PROFUNDO 1480

4 SUR 2372

ESMERALDA 2210

DIABLO REGIMIENTO 2210

Figure 1.1 The topography of the surface and diatreme (Braden pipe) of El Teniente mine. The labels identify the levels mined since the 1940s and those of the future. After more than one hundred years of exploitation, the top of the diatreme now appears in satellite views. (Credit: Felipe Celhay.)

The resulting 90,000 samples are coded by 0 (outside the diatreme) or by 1 (inside). Several approaches have been tested to predict the contour of the pipe:

- *2D approach*: For each drill hole, only the elevation of the transition, either from 0 to 1 or from 1 to 0, is recorded. The regionalized variable is then the elevation of the points on the surface of the diatreme, studied in a traditional way (variogram analysis and then ordinary kriging).
- *Binary approach (3D)*: The indicator function of the diatreme (with values 0 or 1) is directly studied in the three-dimensional space by applying the usual techniques of stationary geostatistics.
- *Intermediate approach (3D)*: The regionalized variable can take the values 0, 1, and 0.5.

Let us detail this third approach. The kriging of a binary random field produces values that are generally included in the interval [0,1], and the question is to know from which value one considers to be inside the object, insofar as the geologists are interested in the coordinates of the surface rather than in a pseudoprobability. In order to answer this question, consider a point in space exactly halfway between two samples coded as 1 and 0, respectively. Because nothing in the model specifies any link between the distance between samples and their values (0 or 1), the midpoint of the two samples will be, on average, on the desired surface. An ordinary kriging at such a midpoint will assign the same weighted ½ to each of the two samples, yielding a prediction equal to 0.5. Accordingly, the value 0.5 constitutes the probability threshold above which it is judicious to consider to be inside the diatreme. The improvement introduced by the third 'intermediate' approach consists in assigning the value 0.5 to the points of the 2D approach that are previously identified as on or close to the surface of the diatreme. The regionalized variable is no longer binary but now has three states: 0, 0.5, and 1.

For each data set, a variogram model is inferred and ordinary kriging is executed. The three approaches are evaluated by cross-validation (prediction of known data values, which are temporarily masked for the exercise) and compared with the work completed twenty years ago by Codelco-El Teniente geologist Felipe Celhay, a Mines ParisTech graduate and coauthor of the publication of the original study (Séguret and Celhay, 2013), who used the same measurements described here, proceeding level by level at a time when geomodeling software did not exist.

Let us analyze the results. For the 2D approach that uses about 1,000 points identified around the surface of the diatreme, some inconsistencies appear (Figure 1.2c and d), such as point n°1 for which the elevation is significantly underestimated. Why? This point is located at an elevation of more than 2,500 m, on the surface, in contact with the air. The transformation to 2D projects it toward the center of the map, in a place where most of the measurements touch a deeper surface whose elevation is obviously less, in the order of 2,000 m. Kriging favors these neighboring measurements and assigns them a significant weight because of their proximity, as nothing in the model indicates that the point targeted for prediction is not of the same nature as its neighbors. In the end, the predicted elevation is too low. This is the price to pay for the simplification that reduces the dimension of the problem. When moving toward the center of the figure, the measurements are supposed to decrease.

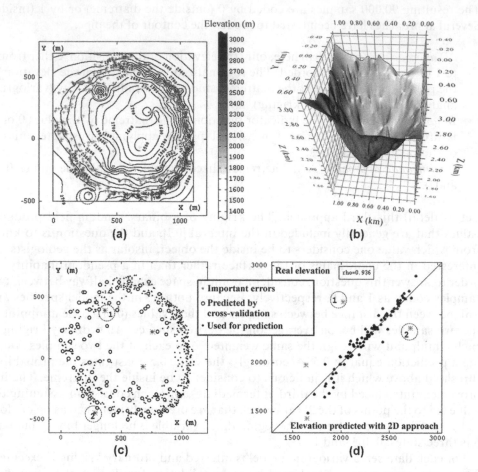

Figure 1.2 2D approach. The predicted diatreme surface represented in (a) isopleths or (b) a vertical development; (c) in light gray the points predicted by cross-validation using the remaining data points; (d) cross-diagram between the prediction (horizontal axis) and the true value (vertical axis). Points n°1 and 2 are discussed in the text.

The same analysis can be done for point n°2, whose elevation is overestimated, in this case because the point is located in the margin of the data crown where essentially high elevations are found. This marginal point is found in a deep protuberence of the diatreme.

For the binary and intermediate approaches, the experimental verification by cross-validation concerns the points of the 2D approach for which kriging is expected to produce 0.5 on average, which is easily observed in the histograms of Figure 1.3: the predictions are centered on this value. Their quality is measured by the standard deviation around the value 0.5. In the intermediate approach, this standard deviation is divided by a factor of almost 2 because, near the surface, there are many data points whose values have been set to 0.5. Their proximity to the point targeted for prediction gives them significant kriging weights.

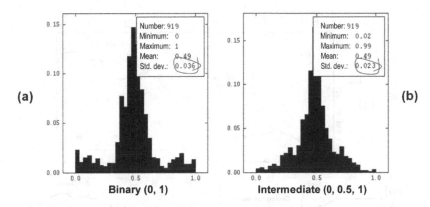

Figure 1.3 Comparison of approaches (1 of 2). (a) Histogram of the predictions at the edge of the diatreme surface using the values 0 and 1 of the binary approach; (b) histogram of the predictions at the same locations using the values 0, 0.5 and 1 of the intermediate approach.

The 0.5 isopleth curves in Figure 1.4a show how these points set to 0.5 'attract' the curve of the intermediate approach. The comparison of the different curves with the geologists' mapping (Figure 1.4b and c) is striking because of the good agreement of the different approaches. The explanation is given in Figure 1.4d, which represents a horizontal section of the result of kriging the indicator data, interpreted as (an estimate of) the probability of being inside the diatreme. In the vicinity of the surface, the thickness of the probability interval [0.25,1] is, on average, about 40 m, a low value in comparison with the dimensions of the diatreme. The sampling information is so abundant that even very different techniques lead to similar results.

The proximity of 2D calculations, which use as few data as approximately 1% of the number of initial samples, with the results of the 3D approach is due to the fact that 80% of the 90,000 initial samples consist of zeros with no influence on the diatreme surface tracing for distances greater than 250 m (diameter of the moving neighborhood used for kriging); the same happens with the remaining 20% of ones located in the center of the diatreme. In the end, near the surface where the methods could provide different results, the number of samples is still favorable to the 3D approach, but by a factor of 10 and not 100. Felipe Celhay, the geologist who initiated this study, says that he prefers the results of the 2D approach, which gives smoother curves that are very close to his hand-drawn layout.

1.1.1 Discussion and perspectives

Initially, this work was, above all, a somewhat provocative exercise aimed at extending the scope of geostatistics to the limit of its possibilities. In fact, the representation of the diatreme as a set of ones in the middle of an infinite space of zeros constitutes the antithesis of any geostatistics due to the nonstationarity of the problem. Indeed, at one time or another, a stationarity hypothesis is required to infer a model (see Appendix, Section A.2.7), which is not possible at the scale of

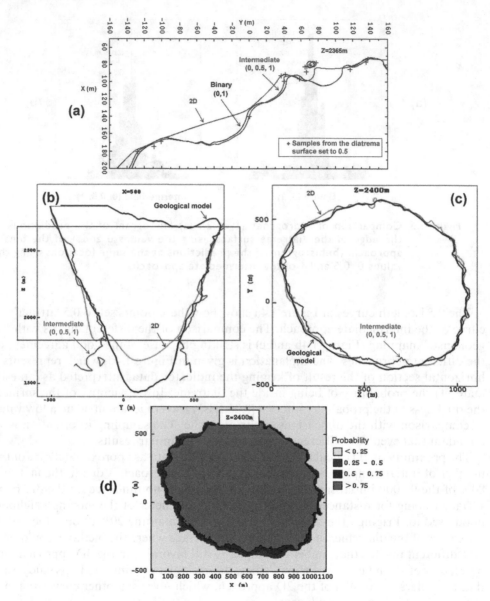

Figure 1.4 Comparison of approaches (2 of 2). (a) Isopleth curves of the intermediate and binary approaches for elevation 2,365 m and surface profile obtained in 2D. The crosses are the known points of the surface, set at 0.5 and that are in the vicinity of 2,365 m elevation. (b) Vertical section of the diatreme, the rectangle represents the study area; (c) horizontal section. At these scales, the binary approach and the intermediate approach are so close that it is impossible to distinguish them. In both cases, the 0.5 isopleth curves are represented. The geologists' layout and the 2D calculations are similar and smoother than the 3D approaches. (d) Horizontal section of the binary kriging representing (estimates of) probabilities of being inside the diatreme.

the diatreme. In the analysis of the prediction results in this unfavorable context, it turns out that, in fact, only the stationarity at the moving neighborhood scale, near the surface of the diatreme, is relevant, with the implicit assumption that the indicator variogram is the same everywhere. In other words, it is assumed that the roughness of the diatreme surface is homogeneous.

The same does not happen with the 2D approach, also due to a stationarity problem. Figure 1.5a presents the variogram of the diatreme surface elevation along the four main directions (N0°E, N45°E, N90°E, N135°E) of the horizontal plane. The behavior is isotropic in the first 150 m, which justifies the use of the average (omnidirectional) variogram of Figure 1.5b. However, the isopleth representation of the diatreme (Figure 1.5c) raises several questions: how is it possible to obtain a linear variogram,

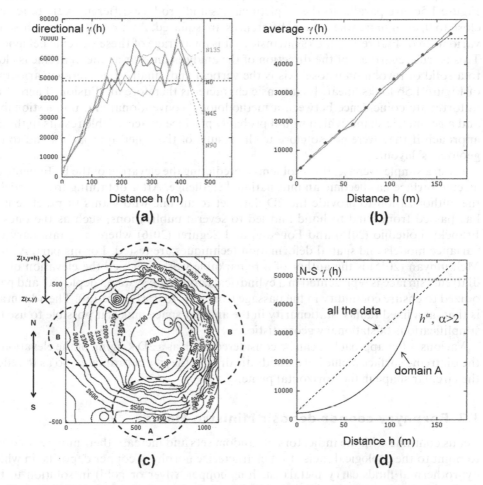

Figure 1.5 Illusory intrinsic stationarity. (a) Directional variograms of the elevations of the 2D approach, along the N0°E, N45°E, N90°E and N135°E directions; (b) omnidirectional variogram; (c) map of the predicted elevations and characteristic domains A and B discussed in the text; (d) power directional variogram (N0°E) obtained when only data pairs in the 'A' domain are used.

indicating stationary increments (a hypothesis known as 'intrinsic' stationarity, see Section A.2.7 in Appendix), in the presence of such strong variations of elevation at small distances?

Take the example of the north–south directional variogram and move in space a couple of points aligned along the north–south direction. When the pair of points is in the 'A' domain of Figure 1.5c, it produces elevation squared differences that give a power variogram with an exponent greater than 2 (Figure 1.5d), a behavior that escapes from the context of intrinsic geostatistics. Given the strong surface gradients at all scales, it would be necessary to use universal kriging (kriging of a random field whose mean value varies in space) or its generalization, the kriging of an intrinsic random field of order k (Chilès and Delfiner, 2012). However, this power behavior does not appear in the results, since it is attenuated by the pairs of points that, taken in the 'B' area of Figure 1.5c, are parallel to the isopleth curves and produce differences that are very close to 0, even more when these differences are squared. As a result, the north-south variogram of Figure 1.5a looks intrinsic as it is an average of these extreme behaviors. This is true regardless of the direction of the analysis, insofar as the diatreme is close to a solid of revolution whose axis is the vertical and, finally, the average variogram of Figure 1.5b looks linear. Its intrinsic character is therefore an illusion. There is an unfortunate coincidence between a methodology, conventional and unquestionable, and a geometric shape, which would probably produce unacceptable results in the 2D approach if they were not so close to the results of the other approaches and to the geologists' layout.

From a simple exercise, this problem of predicting the elevation of the El Teniente diatreme surface has become an international challenge. After obtaining from Codelco the authorization to provide the 2D data set to anyone who wants to practice it, it has passed from hand to hand and led to several publications, such as the ones of Francky Fouedjio (2016) and Fouedjio and Séguret (2016) where nonstationary covariance models and spatial deformation techniques are tested. For his part, Arman Melkumyan (2015) is the author of a remarkable work in which the elevation of the diatreme surface is represented in a cylindrical coordinate system, 'deployed' and periodized to ensure continuity in the passage from 360° to 0° (Figure 1.6). The advantage is that, at a fixed radius, stationarity in the angular space makes it possible to use the simplifications of stationary geostatistics.

Various other approaches can be considered, such as a 2D kriging of the elevation of the diatreme surface including a quadratic drift in the x–y coordinates, so as to reflect the circular shape in the horizontal plane.

1.2 Porphyry copper deposit Ministro Hales

Let us continue with the indicators of random sets and increase their number in order to point to the geological facies that characterize porphyry copper deposits, in which hydrothermal fluids carry metals (such as copper, silver or gold) in solution to the surface through channels of permeability in the host rock, associated with the porosity of the rock and with its discontinuities (cracks, fractures or faults) (Sillitoe, 1998). We leave the Rancagua region and go 2,000 km to the north, in the Atacama Desert, the place of the largest open-pit mines in the world: the legendary Chuquicamata mine, opened in 1882, and its satellites: Radomiro Tomic (in operation since 1997) and

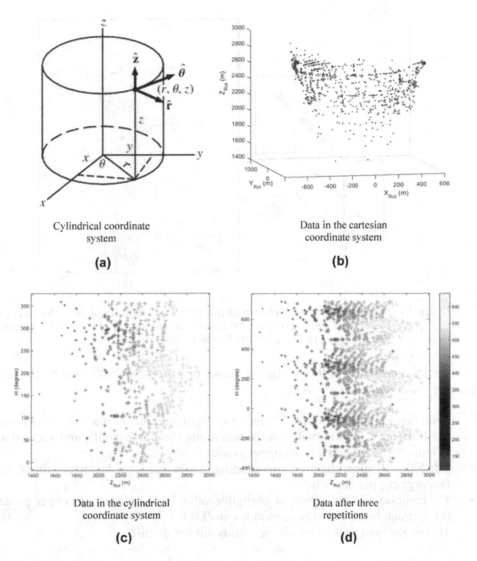

Cylindrical coordinate
system

(a)

Data in the cartesian
coordinate system

(b)

Data in the cylindrical
coordinate system

(c)

Data after three
repetitions

(d)

Figure 1.6 Passage to cylindrical coordinates; (a) principle of the cylindrical coordinates; (b) the known surface elevations are centered around an approximately vertical axis of revolution; (c) data represented in the cylindrical
system; (d) the same representation after a periodization that guarantees
continuity between 360° and 0°. (Credit: Arman Melkumyan.)

Ministro Hales, more prosaically called 'MM', then 'MMH' (in operation since 2010),
which is studied in the following.

On their rise, hydrothermal fluids are, in the MMH deposit, guided by a fault in the
east – the 'MM Fault' or 'West Fault' of the Domeyko fault system – mainly vertical
and north–south oriented (Figure 1.7a), which induces the spatial organization of the

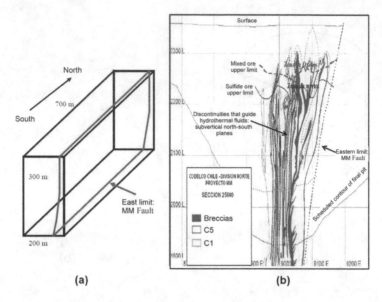

(a)   (b)

Figure 1.7 Ministro Hales. (a) Scheme of the study area and its fault in the east ('MM Fault'); (b) west–east vertical schematic geological section that represents the different facies. (Credit: Pedro Carrasco.)

four facies considered in this study, the names of which are those used by the geologists of the company:

- *Breccias*: highly mineralized breccias of significant economic value, with variable thickness, substantially continuous along the north–south and vertical axes (Figure 1.7b). Their average copper grade is 0.9%.
- *C5*: halo of mineralization surrounding the breccias, of lesser economic value (average copper grade 0.5%).
- *C1*: residual mineralization of negligible value, with an average copper grade of 0.1%, much lower than the economic cutoff 0.4%.
- *Waste*: the host rock in which the metals did not diffuse.

The objective was to question a geological model that, presumably, was giving too much spatial continuity to the mineralized breccias, drawn by hand by joining the known values in the drill holes and, for this purpose, assigning them some uncertainty by using geostatistical simulations. At that time (2003), the mining industry did not have reliable tool to simulate facies, unlike the oil industry for which, for almost fifteen years, the geostatistical team at Mines ParisTech in Fontainebleau, France (at this time, Paris School of Mines), developed the so-called 'truncated Gaussian' method to simulate petrophysical facies (Matheron et al., 1987; Armstrong et al., 2003). A team of three Chilean geologists arrived in France (Pedro Carrasco, his son Pablo and Felipe Ibarra), who, along with us and with Gaëlle Le Loc'h and François Geffroy, but also from Mines ParisTech, were locked up one week in an office to develop what was one of the very first mining applications of this method worldwide.

The method requires specifying several parameters:

- a truncation rule, represented by a two-dimensional or a multidimensional flag, which determines the mutual disposition of the facies and the possible contacts between them;
- a reference system for calculating correlation distances;
- proportion curves that indicate the evolution of the proportion of each facies in space or, more simply, along one direction, often the vertical;
- direct and cross-variograms of the facies indicators.

The truncation rule is indispensable due to the truncation inherent to the method (Beucher et al., 1993). In this study, the rule is imposed by the nature of the mineralization: in the center is the breccia, whose mineralization diffuses in C5, a little in C1, and then stops in the waste. Consequently, the order between facies is given by the sequence breccias–C5–C1–waste, which, in particular, prohibits a direct contact between waste and C5 or between breccias and C1. The evidence of this sequence has been a guide toward this method.

Determining the reference system of the calculations posed a problem: the method was constructed to model sedimentary oil reservoirs, where the vertical direction of sedimentation plays a privileged role in relation to the horizontal, but this is not the case here. Although the solution may seem immediate today, it took us some time before, at the end of the afternoon of the first day, we rotated the computer screen 90° to the right, a screen that is horizontally shown in Figure 1.7b. Victory! The vertical became the east–west direction; there was no longer need to talk of VPC (vertical proportion curves) but of E–WPC. As for the mining equivalent of the stratigraphic horizon from which the proportion curves and variograms are calculated, the digitized surface of the MM Fault is considered for a moment, before the mean plane of the fault is finally adopted, as a vertical plane oriented from north to south (Carrasco et al., 2005, 2007).

To model the proportion curves, it is necessary to gradually move from a single curve (Figure 1.8a) to one curve for each node of the simulated grid, by interpolating the local curves of Figure 1.8b to obtain a result of which Figure 1.8c is a sample. These E–W proportion curves, or the same smoothed curves so as to avoid too large variations of the facies proportions at a small scale, condition the calculation of the direct and cross-variograms of the facies indicators, through the truncation process. The fitting of the resulting variograms was considered satisfactory at that time (Figure 1.9).

1.2.1 Discussion

Compared with Figure 1.7b (hand-drawn layout of the mine geologists), the simulations in Figure 1.10 show a lower continuity of the breccias along the north–south direction, while reproducing the important west–east variability as it is observed in the field. These images of the deposit are not considered more accurate than those given by the geologists but constitute an alternative that allows them to become aware of the tendency, common among modelers, to draw geological objects more continuous than they really are.

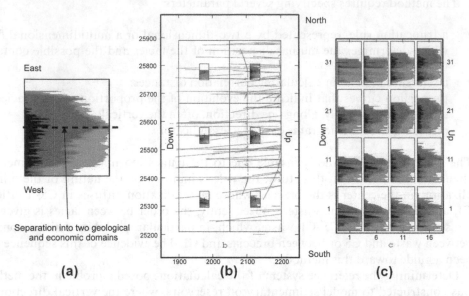

Figure 1.8 Proportion curves. (a) East–west global proportions curve that points up two geological domains; (b) east–west local proportions curves to interpolate at any node targeted for simulation. The lines represent projected drill holes; (c) examples of interpolated east–west proportions curves.

The real gain of this pioneering work was not so much its results as the information collected during the modeling process, and this is a constant throughout this book. In this particular case, it is about the west–east global proportions curves. The dotted line in Figure 1.8a shows a break separating the deposit into two different domains from metallurgical and mining points of view. The western domain contains less C5

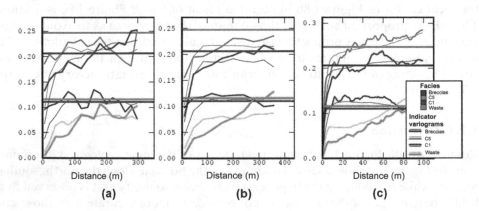

Figure 1.9 Facies indicator variograms. Thick lines: experimental variograms; thin lines: modeled variograms, along the vertical, (b) north–south and (c) east–west directions.

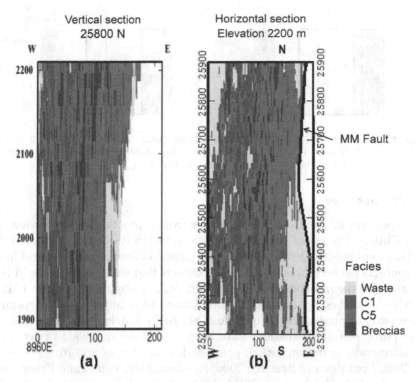

Figure 1.10 Example of facies simulation obtained by the truncated Gaussian method. (a) West–east vertical section; (b) horizontal section.

(rich ore) and more Cl (poor ore) than the eastern domain, in which the proportion of breccias decreases rapidly, while remaining constant in the western domain. Therefore, there is a north–south subvertical limit that, when crossed, will force operators to modify the parameters of the mineral flotation process.

Let us talk about flotation. In this mineral concentration method, the pulp that results from several stages of ore size reduction by crushing and grinding is stirred in cells. The stirring device also injects gas bubbles (see Section 7.1) that make the particles of the minerals of interest rise to the surface, more easily when they are released from their gangue, which is guaranteed if the grinding time is sufficiently long (Fuerstenau et al., 2007). However, this grinding time, the main factor in terms of processing cost, depends on the nature, texture and alteration of the rock containing the minerals. It is not the same for the breccias, C5 or Cl facies, of which the processed ore is a mixture. In mine planning and mineral processing, the boundary indicated in Figure 1.8a is a strong signal of what geometallurgists had expected, something that this study allowed to show and quantify. Subsequently, Codelco requested that a study of the spatial variations of the facies proportions in this deposit be made in order to map this boundary, a work that appears in confidential reports that could one day be made public (Beucher et al., 2008; Séguret et al., 2009).

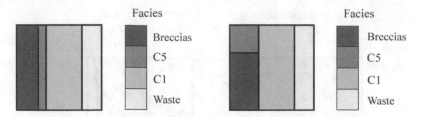

Figure 1.11 Flags representing the truncation rules of the method. On the left, the flag used initially; on the right, the model with two underlying Gaussian random fields. (Credit: Rodrigo Riquelme.)

1.2.2 Perspectives

The perspectives of this work were multiple from a practical point of view. First, in 2008, a Chilean student of Mines ParisTech, Rodrigo Riquelme, improved the truncated Gaussian modeling by introducing a second Gaussian random field in order to allow contacts between breccias and C1, contacts that were not possible in the initial model and that the incorporation of new data made compulsory (Figure 1.11).

Rodrigo used these 'plurigaussian' simulations to quantify the uncertainty of the first five years of production (Riquelme et al., 2008). Lately, numerous other applications of plurigaussian simulations have been carried out in a wide variety of deposit types: diamonds, uranium, nickel, gold, lead, zinc, copper and iron (Deraisme and Field, 2006; Fontaine and Beucher, 2006; Rondon, 2009; Yunsel and Ersoy, 2011, 2013; Talebi et al., 2014; Maleki et al., 2016). An example will be presented in the next section. These facies simulations sometimes incorporate the problem of modeling the grades (Emery and González, 2007; Maleki and Emery, 2015, 2017; Talebi et al., 2016; Mery et al., 2017), which will be discussed in the next chapter.

Recent applications also emerged in the field of geotechnics applied to porphyry deposits, where it is of interest to simulate 'geotechnical units' instead of geological facies, a work contained in a confidential report (Séguret et al., 2016). A geotechnical unit responds to geological and geotechnical criteria, such as the quality of the rock. The use of simulations in this context aims at modeling the uncertainty in regionalized risk factors that are of vital importance for the design of open or underground works. The chapter dedicated to geotechnics will analyze this topic in greater detail.

1.3 Río Blanco-Los Bronces porphyry copper deposit

We will have traveled a lot in this book: in the space of geostatistical methods, in time through works done for more than fifteen years, and in the geographical space since we will not stop traveling Chile from north to south. We leave the Atacama Desert to return to central Chile with a more favorable climate. Operated since 1970, Río Blanco-Los Bronces is another mega porphyry copper deposit located in the Andes mountains, 80 km northeast of Santiago (Serrano et al., 1996; Vargas et al., 1999; Skewes et al., 2003). Currently (2020), the deposit is exploited by open-pit and underground mines operated by the public company Codelco (Andina Division) and the private company AngloAmerican.

Here, the analysis will focus on the portion of the deposit mined by Codelco-Andina, whose geological modeling considers more than 40 rock types that can be grouped into seven main facies. Ordered from the oldest (Upper Cretaceous) to the most recent (Miocene-Pliocene), these are as follows:

- andesite (AND)
- granitoids (GD)
- tourmaline breccia (TOB)
- monolithic breccia (MOB)
- magmatic breccia (MAB)
- porphyry (POR)
- pipe or volcanic chimney-type structure (PIP).

Based on data from several tens of thousands of exploration drill-hole samples (Figure 1.12a), a model has been constructed that provides an interpretation of the geometry of the different facies in blocks of dimensions $15 \times 15 \times 16$ m^3 (Figure 1.12b). Once again, the geologists responsible for resource evaluation would like to know the uncertainty associated with their interpretation.

The chronology of the facies is a guide toward a plurigaussian model based on six random Gaussian fields $Y_1, \ldots Y_6$ and six truncation thresholds $y_1, \ldots y_6$. Once truncated, the first random field defines the volcanic pipe (PIP). In the complement, the second random field defines the porphyry (POR)... and so on until the sixth random field that separates the space not covered by the other facies between the granitoids (GD) and the andesite (AND). This hierarchical modeling allows that a more recent rock (for example, PIP) cross-cuts an older rock. The opposite is not allowed.

The facies data are converted into the indicators of the Gaussian random fields to be truncated (Table 1.1). Variogram analysis of the indicators is then performed, identifying the N20°W, N70°E and vertical directions as the main anisotropy directions. Instead of fitting the indicator variograms, which turns out to be laborious, it was chosen to previously transform them into variograms of the associated random Gaussian fields, by inverting the formula that gives the variogram of a truncated Gaussian field as a function of the variogram of this Gaussian field. The reader will find this key formula, for example, in the book by Chilès and Delfiner (2012). The fitting therefore reduces to modeling six variograms with a unit sill (Figure 1.13).

To finalize the specification of the model, it remains to define the truncation thresholds $y_1, \ldots y_6$. As in the case of the Ministro Hales deposit, the hypothesis of stationarity of the facies is weakened by assuming that these thresholds, and therefore the proportions of the facies, vary in space. In the absence of such a hypothetical reference direction, the interpreted geological model of Figure 1.12b is used. The calculation of the local facies proportions is made from this model. This approach is similar to that used in oil reservoir modeling (Armstrong et al., 2011), except that here a complete geological model replaces the drill-hole data, avoiding the need to interpolate the proportions found in sectors located between drill holes. A moving parallelepiped window of dimensions $135 \times 135 \times 144$ m^3 corresponding to a volume of $9 \times 9 \times 9$ blocks is defined to calculate local facies proportions that continuously evolve in space.

Here again, the plurigaussian simulations (Figure 1.12c–f) illustrate the spatial variability of the facies and their fluctuations around the interpreted geological model.

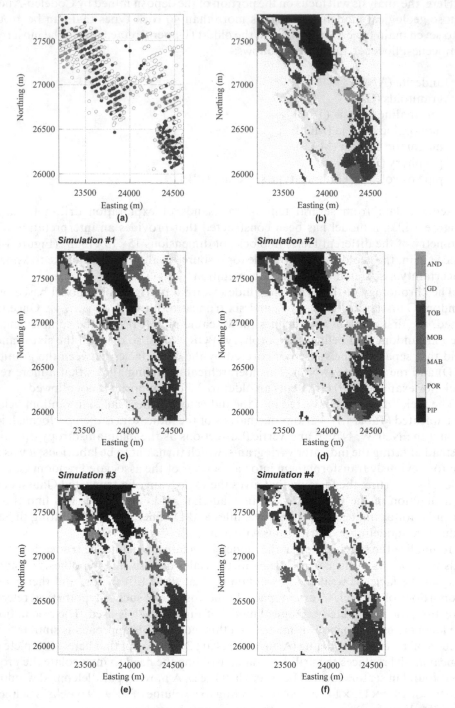

Figure 1.12 (a) Codelco-Andina exploration drill holes; (b) interpreted geological model at elevation 3,348 m, (c)–(f) four conditional facies simulations.

Table 1.1 Facies coding into indicators of random Gaussian fields

Facies	$Y_1 < y_1$?	$Y_2 < y_2$?	$Y_3 < y_3$?	$Y_4 < y_4$?	$Y_5 < y_5$?	$Y_6 < y_6$?
PIP	1	N/A	N/A	N/A	N/A	N/A
POR	0	1	N/A	N/A	N/A	N/A
MAB	0	0	1	N/A	N/A	N/A
MOB	0	0	0	1	N/A	N/A
TOB	0	0	0	0	1	N/A
GD	0	0	0	0	0	1
AND	0	0	0	0	0	0

Each indicator has three possible states: undefined (N/A) when the Gaussian random field does not intervene in the definition of the facies, 1 or 0 otherwise.
AND, andesite; GD, granitoids; MAB, magmatic breccia; MOB, monolithic breccia; PIP, pipe or volcanic chimney-type structure; POR, porphyry; TOB, tourmaline breccia.

Each simulation reproduces the conditioning data, the local facies proportions, the geological continuity reflected in the fitted variograms and the chronological relationships between facies. The volcanic pipe cross-cuts the other facies, whereas the andesite – the oldest facies – does not cross-cut any other facies.

Each simulation is a potential geological model. The set of simulations no longer induces a single deterministic geological model (Figure 1.12b), but a probabilistic model that delivers, for each block, the probability of finding one or another facies (Figure 1.14). In this study, the geological uncertainty is low in the sectors associated with a high probability for some facies, indicating that it is almost certain to find this facies. The transition from the probabilistic model to a deterministic model is still possible, for example, by considering the most probable facies in each block.

1.3.1 Discussion

A critical factor in the implementation of the model is the size of the moving window used to calculate the local facies proportions. A too small window implies that the local proportions in any block are close to 1 or 0, depending on whether the interpreted block corresponds or not to the facies under consideration, and all the simulations become almost identical to the model interpreted by the geologists. Conversely, a too large window produces facies proportions equal to the global proportions in the sector covered by the block model, and in this case, the spatial variations in the proportions disappear: the model is reduced to a stationary plurigaussian model that gives each facies a constant probability of occurrence in space, which is not very satisfactory due to the facies zoning (volcanic pipe to the northern sector, granitoids in the northeast and southwest sectors, porphyry in the central-south sector, etc.). In a way, the size of the moving window quantifies the confidence given to the geologists' interpretation. Cross-validation exercises can be performed to facilitate the selection of this window and to reduce its somewhat arbitrary character (Madani and Emery, 2015).

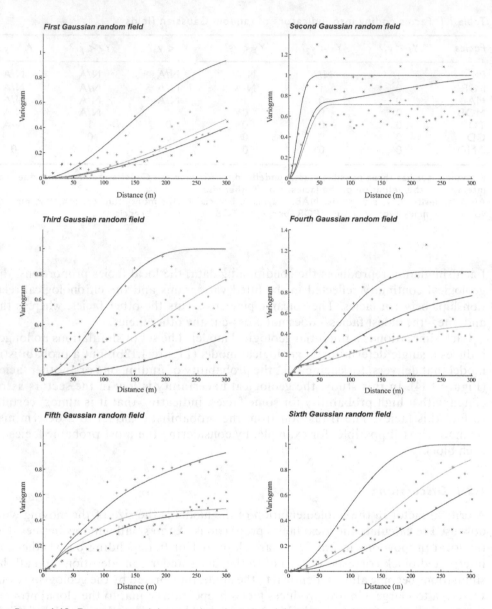

Figure 1.13 Experimental (crosses) and modeled (continuous lines) variograms of the Gaussian random fields to be truncated, along the N20°W, N70°E and vertical directions. The modeled variograms are constructed from nested cubic models with geometric anisotropies.

An alternative to the use of local facies proportions is to model the nonstationarity of the facies through the random fields to be truncated. Madani and Emery (2017) show encouraging results by replacing stationary Gaussian random fields with intrinsic random fields of order k with generalized Gaussian increments, the realizations of which exhibit spatial trends or 'drifts' (see Section A.2.7 in Appendix).

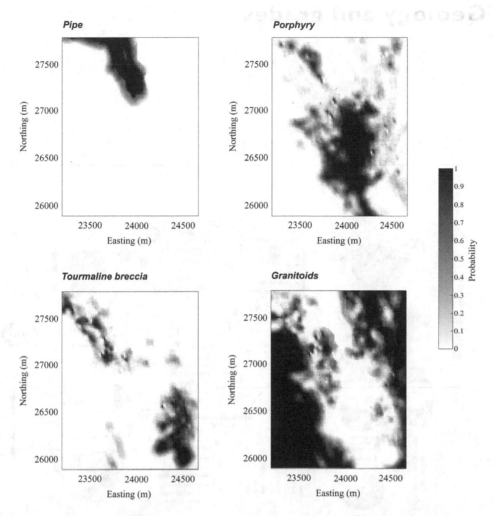

Figure 1.14 Probabilities of occurrence of the four main facies (PIP, POR, TOB, GD), calculated from 100 conditional simulations, for the horizontal section of elevation 3,348 m. GD, granitoids; PIP, pipe or volcanic chimney-type structure; POR, porphyry; TOB, tourmaline breccia.

A final comment refers to the preferential nature of the drill-hole sampling, where the low-grade copper facies – volcanic pipe, porphyry and granitoids – are clearly undersampled (Figure 1.12a). This problem of preferential sampling will be discussed in Chapter 8.

Chapter 2

Geology and grades

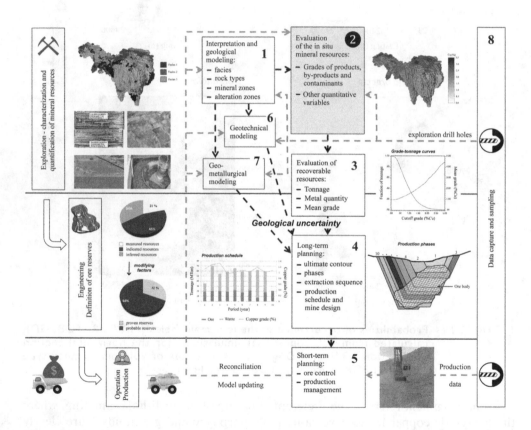

The representation of facies by indicator functions is the first step to solve a broader problem, the construction of a 'block model', that is, the decomposition of a mineral deposit, of several km³, into three-dimensional blocks with dimensions typically between 5 and 20 m on each side (but sometimes smaller or larger). For each block, the grades of elements of interest are predicted or simulated, as well as other features relevant to the extraction or processing of the deposit. This block model will accompany the mining operation from the beginning to the end of its life. It will be, or should be, often modified according to the new information acquired during exploration or production and will condition all planning operations and sizing of works. In a certain way, it represents the direction that a gigantic boat will follow, whose route will be difficult to turn later due to its inertia.

The following sections present two examples of block models made for the Radomiro Tomic (northern Chile) and Río Blanco-Los Bronces (central Chile) porphyry copper deposits. The focus is on geostatistical modeling pertaining to the study and incorporation of the spatial dependence relationships between copper grades and geological facies. In the first example, the formalism of partial grades makes it possible to highlight the absence of edge effects between facies and grades, a property that leads to an original method for predicting the grades. In the second example, the modeling leads to the construction of conditional simulations that reproduce the geological variability, by combining (1) a hierarchical approach in which the geometry of a facies is simulated before its grades in the case where there are no edge effects and (2) a joint approach in which facies and grades are simultaneously simulated, in the case that there are edge effects.

Therefore, this chapter presents the duality between *prediction* and *simulation* prized by geostatisticians on a fundamental mining issue: the construction of a block model that incorporates geology and grades.

2.1 Edge effects and partial grades

The following work refers to the creation of block models, an exciting task for several reasons, including that it allows to fully exercise an important aspect of the role of the geostatistician, which is the role of observer. Our discipline is applied to many fields, and each time it is necessary to empathize with the application domain to understand what is being done; otherwise, the geostatistical calculations may not lead to anything significant or meaningful since the objective, the chosen mathematical means and the interpretation of the results are critical elements.

Here, we draw on our observations of the practices of the geologists of Codelco and other mining companies. Geological facies are modeled first by hand or by using graphic tools (as mentioned in Chapter 1), using objects that are probably too continuous, and then the metal grades are predicted within each facies. This approach results from certain choices, whether conscious or not, explicit or not, such as an implicit hypothesis of independence between the boundary of a facies and the spatial behavior of the grades within this facies. Above all, we have observed that this method produces predictions that are consistent with the production data.

Hence, the approach taken by the geostatistician was to first understand why this practice, based on decades of experience, was working so well. By modeling this practice with the formalism of regionalized variables (Matheron, 1965a, 1971), it may be possible to improve it and to perhaps foresee when it might not work so well. This was the objective, an omnipresent line of conduct in this book.

This role of observer in the problem of grades classified by facies led us to examine closely one of the basic concepts of geostatistics, the search for the 'service variable', the modeling of which can allow the achievement of our objectives. It was about matching the data (of two types, copper grades and facies) to the geostatistical tools, with as few assumptions as possible; in this case, the ones required by ordinary cokriging: the intrinsic stationarity of the random fields associated with the coregionalized variables. The service variable that emerged is the product of a categorical variable (the facies indicator) by a quantitative variable (the grade), as in any bivariate approach. To study the joint behavior of two quantities, the analysis of their average product is often carried out, either statistically through the correlation coefficient or geostatistically through the covariance function or its intrinsic equivalent, the variogram, which measures the correlation between two samples depending on the vector that separates them in the geographical space. This product of a grade by an indicator function will be called a 'partial grade', a name coined by Jean-Paul Chilès, one of whose many qualities is to find the right word.

Once the service variable was defined (or rather, discovered), it remained to unroll a modeling that imposed itself in a natural way. Following the original work of Jacques Rivoirard (1989, 1994), the averages, variances, direct and cross-variograms of the partial grades have a probabilistic interpretation and directly lead to the concept of 'edge effect' or 'border effect'. Its application in several Chilean copper deposits, as well as in a Peruvian zinc deposit and an Australian iron deposit, shows that, quite often, these deposits exhibit a simple statistical property that can greatly simplify the construction of the block model.

The previous statements are now demonstrated by summarizing a work that extended over four years, presented in two conferences (one of mining in Chile and another of geostatistics in Norway) and three publications (Séguret, 2011a, 2013; Séguret et al., 2012). We stay in northern Chile, where the partial grade approach was applied to three deposits:

- Ministro Hales (introduced in the previous chapter);
- 'underground Chuqui', the fabulous resource development project under the current Chuquicamata open-pit mine (Figure 2.1) whose operation is coming to an end;
- Radomiro Tomic, on which the following demonstration will be based. This deposit is currently in operation, and accurate measurements of the blast-hole copper grades have been taken, allowing predictions to be compared with reality.

Figure 2.1 Chuquicamata in July 2018. Open-pit mine in northern Chile, 4 km long, 2 km wide, 1 km deep.

2.1.1 Usual practice

The deposit being divided into parallelepiped blocks, the random field associated with the average metal grade (that is, mass concentration) contained in a block V is denoted as $Z(V)$. The geology is codified by one of n units (facies) at each location x of the deposit where a measurement of the grade $Z(x)$ is available, which is assumed to be at a 'point' support (that is, an infinitely small support).

Geologists consider, on the one hand, the volumetric proportions of the facies and, on the other hand, the grades or mass concentrations. This requires splitting the average block grade as follows:

$$Z(V) = \sum_{i=1}^{n} \frac{v_i}{V} \frac{Q(v_i)}{\rho v_1} = \sum_{i=1}^{n} p_i \, Z(v_i),$$ (2.1)

an expression in which v_i is the volume of the part of V associated with the ith unit, $Q(v_i)$ is the corresponding metal quantity, ρ is the rock density or specific gravity (for sake of simplification, it will be assumed to be identical for all the units), p_i is the volumetric proportion of the ith unit in V and $Z(v_i)$ is the average grade of v_i.

The prediction of $Z(V)$ is made by separately predicting p_i and $Z(v_i)$ and recombining the predictions, here symbolized by the * and ^ signs:

$$\hat{Z}^*(V) = \sum_{i=1}^{n} p_i^* \, \hat{Z}(v_i).$$ (2.2)

The proportion p_i is usually derived from a hand-drawn shape or through the use of software. The geological objects thus produced are intersected with the parallelepiped rectangle V to define the volume v_i. As for the grade $Z(v_i)$, it is predicted by kriging the block-support grade $Z(V)$ using only the grade data belonging to unit i, whose number varies from one unit to another.

2.1.2 Critical analysis

Formula (2.2) is a sum of predictions that does not guarantee the optimality of the result, unless the terms of the sum are in 'intrinsic correlation' (Wackernagel, 2003). Assuming that a cokriging of the summands is used, which would be an improvement to guarantee the consistency of the operations, the number of samples used from one unit to another should be the same, which is often not the case. The units are not evenly distributed everywhere.

On the other hand, each term of the sum (2.2) is a product, the optimality of which is not guaranteed either, even if the terms of the product are spatially independent.

As a last observation, formula (2.1) is complex. Is it really useful to introduce the proportions p_i? Why separate p_i from $Z(v_i)$? What does the product $p_i \, Z(v_i)$ represent? Let us show that this product is the integral, over the support of block V, of the partial grade associated with unit i, the famous 'service variable' that we now introduce.

2.1.3 Reformulation

Let us start by defining the n indicator functions $1_i(x)$:

$$\forall x, i, \; 1_i(x) = \begin{vmatrix} 1 & \text{if } x \in i\text{th unit} \\ 0 & \text{otherwise} \end{vmatrix} \tag{2.3}$$

Every point x in space belongs to a single unit:

$$\forall x, \sum_{i=1}^{n} 1_i(x) = 1, \tag{2.4}$$

a fundamental property on which the following relies. Multiplying Eq. (2.4) by the point-support grade $Z(x)$, one obtains

$$\forall x, Z(x) = Z(x) \sum_{i=1}^{n} 1_i(x) = \sum_{i=1}^{n} Z(x) \, 1_i(x) = \sum_{i=1}^{n} Z_i(x), \tag{2.5}$$

with $Z_i(x)$, the partial grade associated with unit i, defined by

$$\forall x, \; Z_i(x) = Z(x) \, 1_i(x). \tag{2.6}$$

More generally, the total grade is split into a basis of indicators, whatever the volume V:

$$Z(V) = \sum_{i=1}^{n} Z_i(V) \tag{2.7}$$

with

$$Z_i(V) = \frac{1}{V} \int_V Z_i(x) \, dx. \tag{2.8}$$

Assuming that the grade $Z(x)$ and the indicators of the different units are jointly stationary random fields, the partial grades so defined, either on a point support (Eq. 2.6) or on a block support (Eq. 2.8), will also be jointly stationary random fields. Their spatial structure incorporates both the structure of the grade and that of the indicators, something similar to what happens in the formalism of transitive geostatistics when defining the transitive covariogram of a regionalized variable in a bounded domain (Matheron, 1965a, 1971).

The comparison between Eqs. (2.7) and (2.1) shows that the product $p_i Z(v_i)$ is nothing else than the integral over V of the partial grade associated with unit i. It is not useful to separate the terms of the product $p_i Z(v_i)$; its direct prediction by cokriging (CK) will ensure the optimality of the sum:

$$Z(V)^{CK} = \sum_{i=1}^{n} Z_i(V)^{CK}. \tag{2.9}$$

Formula (2.7) leads the problem to an 'isotopic' situation, which is a common concept in geostatistics: at each measurement point, the n partial grades are always known simultaneously. In practice, cokriging is carried out with the n partial grades, and with $n-1$ unit indicators if they are spatially correlated with the partial grades. Looking for a simplification, it is also useful to calculate the cross-variogram $\gamma_{i Z_i}$ between the indicator and the partial grade of unit i, divided by the indicator direct variogram γ_i. Under a hypothesis of spatial symmetry (Séguret, 2013), this quotient reflects the average behavior of the grade when, from the outside of the unit, one proceeds to move inside:

$$\left| \frac{\gamma_{i Z_i}(h)}{\gamma_i(h)} \right| = E\{Z(x+h) \mid x+h \in i, x \notin i\}, \tag{2.10}$$

the famous 'edge effect' (Rivoirard, 1994). It should be noted that it is possible to get rid of the spatial symmetry hypothesis, by substituting noncentered covariances for the direct and cross-variograms, which would allow to detect a directional edge effect (expectation of $Z(x+h)$ when $x+h \in i$ and $x \notin i$ different from the expectation of $Z(x)$ when $x+h \notin i$ and $x \in i$).

If the quotient (2.10) does not depend on the separation vector h, the following notable property arises:

$$Z_i(x) = m_{/i} 1_i(x) + R_i(x), \tag{2.11}$$

with $m_{/i}$ the global average of the grades belonging to unit i. This equation expresses the partial grade as a linear combination of the indicator 1_i with a residual R_i that is spatially uncorrelated with this indicator.

If, in addition, all the residuals do not present any spatial cross-correlation and are spatially uncorrelated with the indicators of the different units, then the following simplification takes place:

$$Z(V)^{CK} = \sum_{i=1}^{n} \left[m_{/i} 1_i(V)^{CK} + R_i(V)^K \right], \tag{2.12}$$

where $1_i(V)$ and $R_i(V)$ represent the average values of $1_i(x)$ and $R_i(x)$ over block V, while the abbreviations K and CK placed in superscript stand for kriging and cokriging, respectively.

Cokriging each partial grade then reduces to cokriging $n-1$ indicators and separately kriging the residuals. The calculations are accurate and do not suffer from nonoptimal approximations as for Eq. (2.2). The property (2.11) also explains why the usual practice of contouring the geological units before calculating the grades in each unit does not give bad results, insofar as it establishes the absence of edge effects and, consequently, the independence between the categorical variable (the spatial layout of the geological unit) and the behavior of the grades within the unit. This formula also states that, as a first approximation, it is sufficient to assign, to each unit so delineated, the global average grade associated with this unit, a procedure that will not produce any bias. At worst, the precision, quantified by the prediction error variance, will not be very good.

This property also allows obtaining a procedure for simulating the grades by taking into account the specificity of the geological units or facies. For the facies and their indicators, it has been seen in the previous chapter that the truncated Gaussian (TG) or the plurigaussian (PG) method is appropriate. For the residuals, a simulation of a transformed Gaussian random field is possible, for example, by the turning-bands (TB) algorithm (Matheron, 1973; Emery and Lantuéjoul, 2006). One then has:

$$Z(V)^{\text{simu}} = \sum_{i=1}^{n} \left[m_{/i} 1_i (V)^{\text{simu}\_PG} + R_i (V)^{\text{simu}\_TB} \right]. \tag{2.13}$$

This amounts to simulating the partial grade Z_i (with a deterministic mean $m_{/i}$ and a random residual R_i) in the simulated facies associated with this partial grade. Such a hierarchical method is often used when simulating facies and grades.

The property (2.11) has been verified in a Peruvian zinc deposit (Candelario, 2010), as well as in the Ministro Hales and underground Chuquicamata deposits. Next, it is shown that it is also true for Radomiro Tomic.

2.1.4 Application: Radomiro Tomic

We describe here a study completed by the authors using about 25,000 drill-hole samples of length 3 m, with information on the copper grade and a classification among one of the following four facies: veins (equivalent to the previous MMH breccias, very rich), C5 (rich), C2 (equivalent to the previous C1, not very rich) and waste. The mineralized facies are present throughout the entire domain under study.

When distinguishing by facies (Figure 2.2), the copper grades range from the poorest (waste, average grade 0.16%) to the richest (veins, average 1.33%). The range of values is small, with an average of 0.36% for C2 and 0.67% for C5. These differences are important for the project economics, given the size of the deposit (several km^3), but numerically, the various averages fluctuate in a small range compared, for example, with Peruvian zinc deposits (where grades could be multiplied by a factor 10) or iron deposits in Brazil or Australia whose grades reach up to 70%. The histograms in Figure 2.2 further show that the facies share fairly wide ranges of grades. The facies are not determined by the grades and constitute genuine information, complementary to the grade information.

The evaluation of the edge effects (Eq. 2.10) indicates a total absence of these effects for three facies and a not very significant variation for the veins: the average grade increases by less than 0.1% between the edge and the center of the veins, a variation that does not justify complicating the calculations (Figure 2.3).

Here again, although in general such variations are important, it is worth remembering that cokriging will be used with a moving neighborhood, at a scale of approximately 100 m, and it should be determined whether the cokriging system can be simplified. Such a simplification is possible here, the property (2.11) applies and cokriging four partial grades with three out of the four indicators reduces to cokriging the indicators and kriging the residuals (Eq. 2.12). Beforehand, it is verified that the residuals have no spatial cross-correlation and are also spatially uncorrelated with the indicators of the different units.

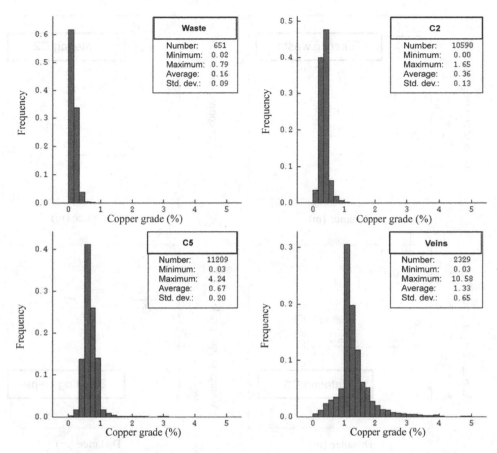

Figure 2.2 Histograms of copper grades by facies.

Production data, in the form of blast-hole measurements, are available in a part of the domain under study that has been in operation for several years. These blast holes can be considered as representing the reality because they are in a five times finer mesh than the drill-hole samples and are located in and around each block. Determining the destination of the blasted material – to the processing plant or waste dump – is decided on the basis of an ultimate prediction using the blast-hole measurements. These measurements differ from the diamond drill-hole samples – the question arises about the consistency of their grades with those measured on drill holes; this question is the subject of Chapter 5, with a favorable conclusion for the blast holes. For now, we follow the mining geologists in their almost universal practice: consider the blast holes as a reference information and the prediction of the block grade based on them as the 'reality'.

When the values obtained by cokriging the partial grades are compared with the reality, the correlation is 0.69 (Figure 2.4c), significantly better than 0.63 of the model constructed by the geologists (Figure 2.4b), due to a group of high grades underestimated by the geologists but better predicted in the case of cokriging. Apparently,

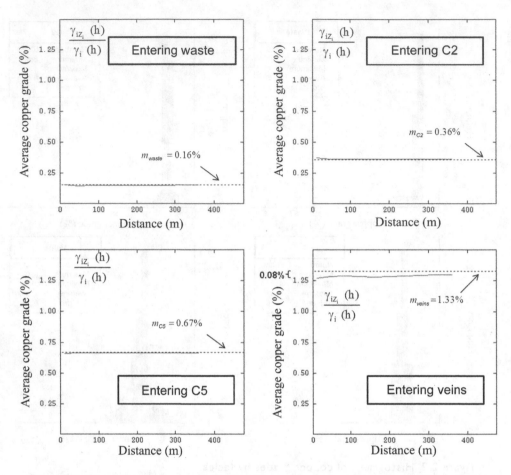

Figure 2.3 Edge effects. For each facies, the cross-variogram between the indicator and the partial grade divided by the indicator direct variogram is represented. This ratio indicates the presence or absence of an edge effect. When a horizontal line is obtained, there is no dependence relationship between the expected grade at an interior point of the facies and the distance that separates this point from the facies boundary. This is the case here of three facies, the fourth one (veins) being the only one that has a weak edge effect (less than 0.1% enrichment between the edge and the center).

the geologists' interpretation was conservative and included some isolated pieces of very rich facies inside poorer facies, whereas the geostatistical approach assigns each data a weight that depends on its distance to the block targeted for prediction. The correlation between the two approaches is surprisingly low, less than 0.7 (Figure 2.4a). Figure 2.4d represents a 3D scatter plot between the two approaches and the true grades, which are somewhere between the two predictions.

In what do these two methods really differ? The property (2.11) shows that, in relation to a possible bias, the decisive step is the delineation of the facies. On the one

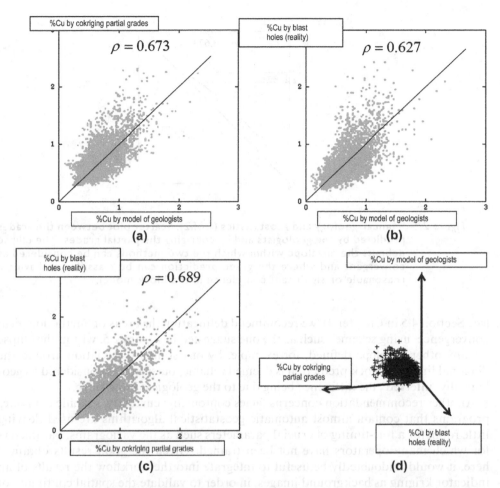

Figure 2.4 Predictions and reality. (a) Scatter plot between the grades predicted by the geologists and by cokriging the partial grades. (b) Scatter plot between the grades predicted by the geologists and the true grades deduced from the blast holes. (c) Scatter plot between the grades predicted by cokriging the partial grades and the true grades deduced from the blast holes. (d) 3D scatter plot between the predictions with two methods and the true blast-hole grades. The partial grades method gives better results. Its implementation takes two weeks, whereas the geological model occupies three geologists for more than one year.

hand, they are drawn by hand by geologists who introduce their interpretation and knowledge of the geological processes; on the other hand, the facies result from an indicator cokriging that only aims at unbiasedness and optimality. One method represents knowledge, while the other one represents objectivity. In the end, both methods seem indispensable, which is the true conclusion: one must combine the geostatistical quantification of uncertainty with geological knowledge. For example, in the resource classification stage that aims at giving a prediction reliability index for each block

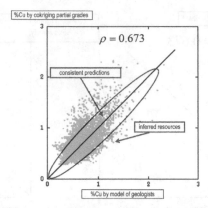

Figure 2.5 'Manual' geology and geostatistics (1 of 2). Scatter plot between the grades predicted by the geologists and by cokriging the partial grades. The ellipse delimits the envelope within which the two methods can be considered as convergent and where the grade prediction can be classified as having a reasonable or significant confidence in the block model.

(see Section 4.5 in Chapter 4), we recommend delineating the areas of 'methodological convergence' using schemes such as the one suggested in Figure 2.5: within the ellipse (or any other envelope defined, for example, by one standard deviation around the diagonal line), the block prediction is deemed reliable; outside, it is considered as geologically 'inferred', the difference being due to the geological knowledge.

Another recommendation concerns facies contouring: currently, graphic computer programs that contain almost automatic geostatistical algorithms are used, leaving little room for a fine-tuning of crucial parameters such as the variograms, parameters for which many operators have not been trained. Given the good results obtained here, it would undoubtedly be useful to integrate into the workflow the results of an indicator kriging as background images, in order to validate the spatial continuity of the structures identified in the drill holes (Figure 2.6).

2.1.5 Discussion and perspectives

The construction of a block model by cokriging the partial grades in the Radomiro Tomic deposit has demonstrated the main potential of the partial grades method: it complements the geologists' approach, and two examples have been given to show how the two approaches can be combined.

In contrast, in the underground Chuquicamata project, which also exhibits the property (2.11), the partial grades approach was compared with a 'blind' kriging that ignores the facies, and it was found, by cross-validation based on 10% of the measurements, that the differences were not significant. Three reasons explain this result:

* On average, the facies grades are not sufficiently different.
* The ranges of the indicator residual variograms are similar to the ranges of the indicator variograms, which induces a kind of redundancy in the model.

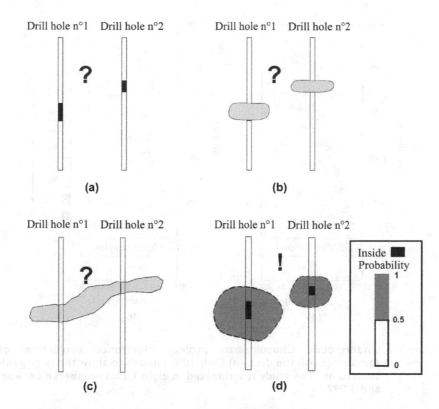

Figure 2.6 'Manual' geology and geostatistics (2 of 2). The geologists' layout could be guided by geostatistics. (a) Initial data, facies in black. (b) Two objects or (c) a single object? (d) Map of the probability of being in the facies obtained by cokriging the facies indicators. The dotted contour represents the probability 0.5. Here, the two contours do not touch; the probability of a union between the two objects is low.

- At the scale of the cokriging moving neighborhood, only two or three of the eight facies of the underground project are present, as the facies are grouped into large geographical zones (Figure 2.7a).

Similar synthetic deposits were simulated with more contrasted grades, reproducing the behavior of the facies. Focusing on zones where many facies are simultaneously present at the scale of the production blocks, it is possible to show that the results are a little better, but not by much (Séguret, 2013). The main reason is conceptual: cokriging allows the joint use of different types of measurements, and the practice demonstrates that it is more effective when it involves measurements known at different locations ('hererotopic' situation). But here, the cokriging of partial grades is 'isotopic' and does not allow to integrate in the system any new spatial positions with a measurement of either a primary or a secondary variable. The improvement is only due to the model, not to the spatial sampling, which reduces the efficiency of cokriging.

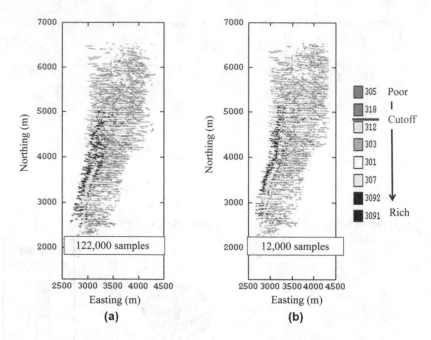

Figure 2.7 Underground Chuquicamata project. Horizontal projections of the samples: (a) All the data. (b) Only 10%, enough to show the large geological structures. The study is interested in eight facies numbered between 301 and 3,092.

The benefit of this work is not so much in the results of the prediction obtained, as in the properties discovered throughout the modeling, being in the first place the property (2.11) of independence between the facies contours and the grades within the facies, which leads to a new problem in geostatistics that could be the subject of a doctoral study. Let us look at Figure 2.7a and b.

In both cases, the main geological structures drawn by the facies are clear, although the figure on the right contains less than 10% of the samples in the figure on the left, which raises the question: why have so many samples been taken, knowing that a meter of drilling costs several hundred dollars? The answer is that the successive drilling campaigns of this project are oriented toward the increasingly fine recognition of the grades by volume, and there is always a need for more drilling to capture mineralized facies that vary in size from several meters to the thickness of a hair. However, the previous study shows that the economics of the project is determined by the contours of the geological units (more so than the grades themselves), which can be delineated with considerably fewer samples than those available. By focusing the sampling campaigns on the geological unit identification and not on the grades, it would have certainly been possible to achieve significant savings. Furthermore, unless very specific geometric shapes are considered, sampling at regular intervals the surface surrounding a volume requires fewer samples than sampling the entire volume with the same sampling density. Then another question follows: how could geostatistics guide the sampling campaign for this type of predominantly geometric deposit? The answer to this problem is not immediate and should be the object of future developments.

Let us conclude this section with the analysis of another tool obtained from the modeling: the 'preferential contact scheme'. While the mutual behavior of grades and facies has been the focus of our attention so far, we should also mention the mutual behavior of the facies themselves and the analysis of their transitions. The principle of the analysis of such transitions appears in a course written by Jacques Rivoirard in 1990 and published in 1994. As for the first applications, one must go back to the work of sedimentologist Walter Schwarzacher, who, in 1969, in the context of Markov chains, began a work followed later by hydrogeologists Gary Weissmann and Graham Fogg in 1999, making the link with geostatistics. Later, in 2006 and 2007, Weidong Li specialized in the analysis of categorical variable transitions and introduced the term *transiogram*.

Similar to formula (2.10) where the cross-variogram between an indicator and a partial grade is divided by the indicator direct variogram to obtain a conditional expectation, it is possible to replace the partial grade with another indicator and thus define the quotient:

$$\left|\frac{\gamma_{ij}(h)}{\gamma_i(h)}\right| = \text{Prob}\{x+h \in j \mid x \in i, x+h \notin i\}, \tag{2.14}$$

which, under a hypothesis of spatial symmetry (Rivoirard, 1994), represents the probability of entering facies j when leaving facies i, as a function of the vector h that separates the two points at which this probability is calculated. When the direct and cross-variograms involved in Eq. (2.14) have a sill, the quotient in the formula (2.14) tends to a sill equal to $p_j/(1 - p_i)$, the probability of being in facies j divided by the probability of not being in facies i, two probabilities whose calculation is based on the global facies proportions p_i and p_j.

The quotient (2.14) is useful to rank the facies. If $|\gamma_{ij}(h)/\gamma_j(h)|$ depends on vector h, but $|\gamma_{ij}(h)/\gamma_i(h)|$ does not, then facies i has, in some way, priority over facies j, whose indicator can be expressed as a linear combination of the indicator of facies i and a spatially uncorrelated residual:

$$1_j(x) = \frac{p_j}{p_i - 1}\left(1_i(x) - 1\right) + R_{ji}(x). \tag{2.15}$$

This property implies that cokriging the indicators reduces to the separate kriging of the indicator and the residual. If such properties are met for several pairs of facies, the linear cokriging system of facies indicators can be greatly simplified. The following example presents an original use of the indicator cross-to-direct variogram ratios for the Chuquicamata project.

The curve starts from the value $p(\rightarrow j | i \rightarrow)$, for h close to 0, which represents the probability of finding facies j when leaving facies i, a probability in practice evaluated by the total number of pairs of samples of facies i and j that are directly in contact, divided by the total number of samples of facies i. The goal is to detect preferential contacts, i.e., a tendency, for the facies, to meet some more than others, the idea being to

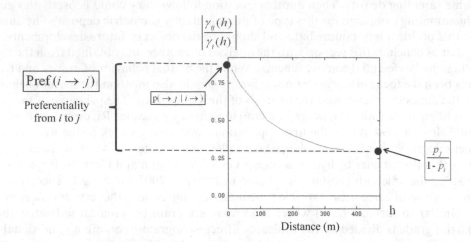

Figure 2.8   Quotient of an indicator cross-variogram by an indicator direct variogram
for the underground Chuquicamata project. Text contains explanations.

generalize the concept of contact diagram (flag) mentioned in Chapter 1 (Figure 1.11). However, as shown in Figure 2.8, the value $p(\to j|i \to)$ is not interpretable in itself as it must be relativized by the value to which the indicator cross-to-direct variogram ratio converges (when it converges), that is, the probability ratio $p_j/(1 - p_i)$. Hence, one must introduce the following quantity called 'preferentiality from i to j':

$$\text{Pref}(i \to j) = p(\to j \,|\, i \to) - \frac{p_j}{1 - p_i}, \tag{2.16}$$

which represents the 'probability jump' between the probability of direct contact and the probability related to the global facies proportions. This quantity depends on the direction of the vector h joining the two points. In practice, it is calculated in the east–west, north–south and vertical directions. It should be taken into account that using a difference in Eq. (2.16) allows preserving the interpretation in terms of probability, which a quotient between $p(\to j|i \to)$ and $p_j/(1 - p_i)$, for example, would not allow. Some rules of interpretation of this quantity are found in Séguret (2013).

For Chuquicamata, in each of the three directions, all the preferentialities between the eight facies of the deposit (56 values per direction) were calculated and classified by order of importance. The pattern that appeared is almost the same for all the directions (Figure 2.9). To understand this scheme that was emerging from statistics, an impressive global law established by measuring the relationships between facies taken two by two at the scale of the drill hole samples, we had to discuss with the resources geologist Pablo Carrasco. The remembrance of his reaction is still alive when, in December 2011, he discovered this preferential contact pattern: he examined it carefully, silently, with a visible emotion on his face. It is good to specify here that Pablo is the son of Pedro Carrasco, who, as stated in the introduction, was an initiator of our collaboration with Codelco and died of illness one year before his son explained what he had in front of our eyes.

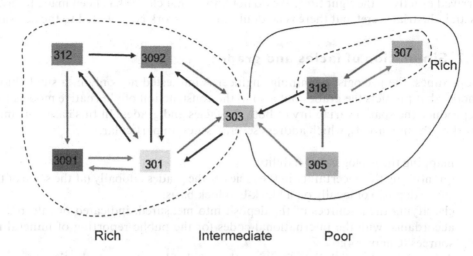

Rich   Intermediate   Poor

Figure 2.9 Diagram of preferential contacts between the underground Chuquicamata units. The arrows can be colored to indicate the importance of the transition probabilities between two units. Therefore, when one leaves 312, one mostly finds 301, whereas when leaving 301, one finds almost as much 3,091 as 303, and sometimes 3,092. 312 and 301 do not have a symmetrical connection, 301 screens 312 with respect to the other units, which establishes an order relationship between these two units. When two arrows appear between two units, there is a mixture and there is no longer an order, and none of the two units screens the other with respect to the other units. Economically speaking, the challenge of the project here is unit 307 at the east end of the scheme. Should it be exploited, given that it would require mining the metal-poor unit 318?

First, the diagram must be oriented as in Figure 2.9, west to the left and east to the right. The facies are grouped into two sets, separated by facies 303, which was called 'transition facies' twenty years before – how curious… The association of these groups with the grades shows that, if the entire west is a mixture of four units whose contours are probably quite difficult to delineate, the grade of this 'mixture' (in the sense of the large number of double arrows between facies) remains above the economic cutoff grade. In any case, they will generate profits after extraction, whether or not their individual contours are accurate, especially since the exploitation by block caving is not at all selective and must mine all the ore in a given area, several hundred meters below the present Chuquicamata open pit (itself 1,000 m deep). In the east, the facies highlighted by the scheme are much less favorable for project economics. To get to unit 307, one must cross unit 318 and avoid unit 305 that contains little metal.

At the time this work was presented to the company, the crucial question was whether the exploitation should stop at the eastern end of unit 303, or if it was necessary or beneficial to take the exploitation to unit 307, at the price of extracting a large amount of low-grade ore (unit 318). Geologists and engineers found in this scheme the synopsized, qualitative and quantitative expression of a truth of which they had the intuition for twenty years. They also obtained a method to model the blocks and to better identify the rich areas of the deposit when surrounded by poor ore. These results arrived exactly at the right time. We do not know what choice has been made today, it is an industrial secret, but there is no doubt that this work contributed to the decision.

2.2 Simulation of facies and grades

Sometimes, the geologist or mining engineer is interested not only in a single block model that predicts the grades but also in the construction of alternative models that reproduce the spatial variability of both the facies and grades. In this case, one must work with simulations, which address several issues, in particular:

* mapping the geological variability;
* quantifying the uncertainty in the facies or the grades, globally (at the scale of the entire deposit) or locally on a block-by-block basis;
* classifying the resources of the deposit into measured, indicated or inferred, in accordance with the international codes for the public reporting of mineral resources (Chapter 4);
* determining the reliability of mining plans and of long-term production programs (Chapter 4);
* assisting the ore control function and short-term production management (Chapter 5);
* defining a more detailed sampling plan to convert inferred resources (with low confidence) into indicated or measured resources.

The work described in the following by Emery and Maleki (2019) presents a case study of facies and grade simulation in the Río Blanco-Los Bronces porphyry deposit introduced in Chapter 1. First, the facies are modeled using two truncated Gaussian random fields (plurigaussian model). The relationships between facies and copper grades are then examined in light of three exploratory tools, revealing the coexistence of facies with and without edge effects. A four-variable coregionalization model is then developed to describe the joint spatial correlation structure of the facies and grades. Finally, a hierarchical approach is used to simulate one facies and its grade, whereas the other facies are cosimulated together with their grades. The results are cross-validated by the so-called 'split-sample' technique.

2.2.1 Data and sector under study

Given the large spatial extent of the Río Blanco-Los Bronces porphyry deposit, this study focuses on a small sector mined by open pit and known as Don Luis, with a volume of $600 \times 775 \times 450$ m^3 (Figure 2.10). The available data include more than 3,000 samples of exploration drilling, composited at a length of 16 m, with information on the rock type and the in situ copper grade (Table 2.1). The rock type belongs to three of the seven main facies recognized in Chapter 1 (the other four facies being absent from the sector under study):

* granitoids (GD);
* mineralized tourmaline breccia (TOB);
* porphyry (POR).

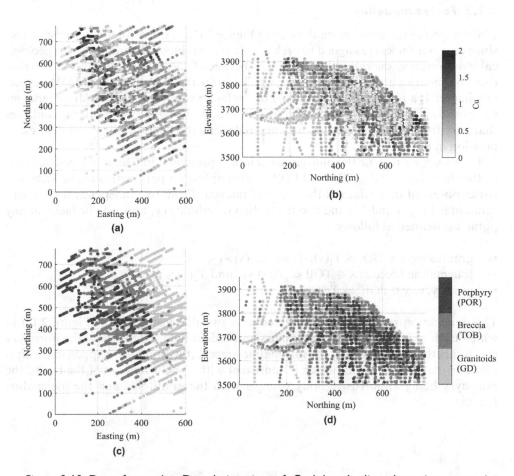

Figure 2.10 Data from the Don Luis mine of Codelco-Andina. Location maps in local coordinates: projections in horizontal (a, c) and vertical sections (b, d). The scale indicates the copper grade (a, b) and the geological facies (c, d).

Table 2.1 Statistics of in situ copper grade data, throughout the sector under study and in each facies

	Entire study area	Granitoids	Tourmaline breccia	Porphyry
Number of data	3,085	1,288	1,351	446
Minimum	0.02	0.04	0.03	0.02
Average	0.69	0.64	0.89	0.25
Maximum	3.45	3.45	3.32	1.26
Variance	0.158	0.097	0.152	0.031

Copper grades are expressed in %.

2.2.2 Facies modeling

Codelco geologists have drawn the map in Figure 2.11, whose resolution is quite coarse since only one facies is assigned to each $15 \times 15 \times 16\,\mathrm{m}^3$ block. In addition, this geological interpretation smoothes the boundaries between facies and implicitly assumes that these boundaries are perfectly known, which is unrealistic. As with the MMH deposit in Chapter 1, a plurigaussian modeling of the facies would avoid these disadvantages by providing a set of high-resolution simulations, with spatially irregular boundaries that vary from one simulation to another, reflecting the uncertainty about their exact position in the deposit.

Therefore, we resume the plurigaussian model presented in Chapter 1, restricting it to the three facies (GD, TOB and POR) and adapting its parameters to the local statistics observed in the data of the sector of interest. From two independent Gaussian random fields (Y_1 and Y_2) and two truncation thresholds (y_1 and y_2), the facies at any point x is defined as follows:

- granitoids: $x \in \mathrm{GD} \Leftrightarrow Y_1(x) > y_1$ and $Y_2(x) > y_2$,
- tourmaline breccia: $x \in \mathrm{TOB} \Leftrightarrow Y_1(x) > y_1$ and $Y_2(x) \leq y_2$,
- porphyry: $x \in \mathrm{POR} \Leftrightarrow Y_1(x) \leq y_1$.

This truncation rule can be represented as a two-dimensional flag (Figure 2.12) in which each axis corresponds to a Gaussian random field. In this flag, the porphyry cross-cuts the granitoids and the tourmaline breccia (the same will happen in the simulations that will be built), which is consistent with the chronology of the facies, the porphyry being an intrusive body younger than the granitoids and the tourmaline breccia.

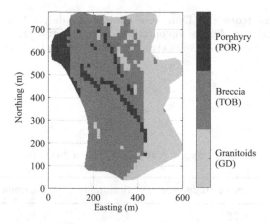

Figure 2.11 Interpreted geological model of the Don Luis mine of Codelco-Andina, with a block size of $15 \times 15 \times 16\,\mathrm{m}^3$. Horizontal projection of the bench with elevation 3,650 m, restricted to the sampled domain. Granitoids are concentrated in the east and porphyry in the west and center of this domain. (Credit: Claudio Martínez.)

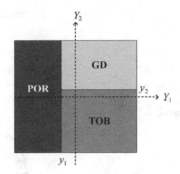

Figure 2.12 Flag that symbolizes the plurigaussian truncation rule, which allows obtaining the facies (GD, TOB or POR) from the values of the Gaussian random fields Y_1 and Y_2. GD, granitoids; POR, porphyry; TOB, mineralized
tourmaline breccia.

The truncation thresholds $y_1 = -0.8$ and $y_2 = 0.03$ are set in order to reproduce the
global proportions of the facies in the interpreted geological model for the sector under study (38.5% for GD, 40.3% for TOB and 21.2% for POR). The variograms are
fitted hierarchically, considering the plurigaussian model as two embedded truncated
Gaussian models:

- the direct variogram of the Gaussian random field Y_1 is modeled first, in order
  to fit the experimental variogram of the porphyry indicator (POR = 1, GD = 0,
  TOB = 0) (Table 2.2 and Figure 2.13a);
- next, the direct variogram of the Gaussian random field Y_2 is modeled, by fitting
  the experimental variogram of the tourmaline breccia indicator defined in the
  space not occupied by the porphyry facies (TOB = 1, GD = 0, POR = undefined)
  (Table 2.2 and Figure 2.13b).

The direct variogram models of Y_1 and Y_2 comprise nested spherical structures having
a geometric anisotropy with main directions N20°W, N70°E and vertical, the same
directions as the ones identified in Chapter 1. In addition, it is assumed that these two
random fields are independent, so that their cross-variogram is identically zero.

Table 2.2 Facies coding by means of three-state indicators

Facies	$Y_1 < y_1$?	$Y_2 < y_2$?
POR	1	N/A
TOB	0	0
GD	0	0

States 1 and 0 correspond to the cases when the underlying Gaussian ran-
dom field is lower or higher than the truncation threshold, respectively.
The undefined state N/A corresponds to an absence of information on the
Gaussian random field (missing indicator data).
GD, granitoids; POR, porphyry; TOB, mineralized tourmaline breccia.

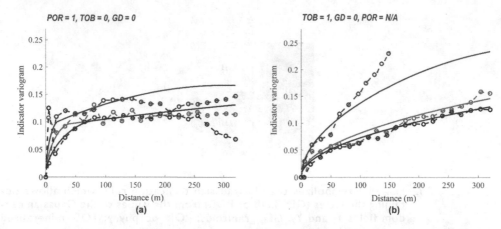

Figure 2.13 Fitting of the direct variograms of facies indicators: (a) porphyry (POR), (b) tourmaline breccia (TOB). Experimental variograms (circles and dotted lines) and modeled (solid lines) are calculated along the N20°W, N70°E and vertical directions, identified as the main directions of anisotropy. GD, granitoids.

2.2.3 Study of edge effects in the copper grades

Before modeling the grades within the facies, it is important to determine whether or not these grades exhibit an edge effect when moving from one facies to another. Two extreme situations are possible:

- Absence of edge effects: The average copper grade remains constant within each facies, changing abruptly, or presenting a very rapid transition, when crossing the boundary between two facies.
- Presence of an edge effect: The average copper grade within a facies varies when approaching the boundary and may, in the extreme case, not present any discontinuity when passing to an adjacent facies.

In the first case, as explained in Section 2.1 on the formalism of partial grades, it would be advisable to simulate the grades in each facies separately (Eq. 2.13). In the second case, however, the same random field could be used to simulate the grades in several facies, so as to avoid discontinuities when moving from one facies to another. This unique random field can also be correlated with the facies indicators, because the presence of an edge effect does not necessarily mean that the grade is independent of the facies.

Accordingly, there are three possible approaches to simulation:

- A hierarchical approach, where facies are simulated first and then the grade is simulated in each facies separately.
- A joint approach, where facies and grades are simulated simultaneously. In this case, the aim is to cosimulate a quantitative variable and a categorical variable,

a problem that ultimately reduces to cosimulating Gaussian random fields conditioned to exact data (copper grades) and interval data (indicators associated with truncation thresholds). The conditioning step is solved by using the Gibbs sampler, an iterative technique that transforms interval data into Gaussian data (Emery and Silva, 2009; Maleki and Emery, 2015).

- A mixed approach, in the case that the copper grade does not have an edge effect with given facies, but presents an edge effect with other facies, an approach that will be preferred in the present case study.

The ratio of the cross-variogram of the partial grade and the facies indicator by the direct variogram of the indicator (Eq. 2.10) is used to determine the existence or not of edge effects. It gives the average grade in the facies as a function of the distance to the boundary. Applied sequentially to each facies, this ratio does not reveal any edge effect, as the average grades remain constant within each facies, regardless of whether one approaches or moves away from the boundary with another facies. The average copper grade in the porphyry fluctuates around 0.25% (Figure 2.14c), which is significantly lower than the average grade in the granitoids (Figure 2.14a) and tourmaline breccia (Figure 2.14b). These last two facies exhibit relatively close average grades (0.8%–0.9%) near their boundaries.

Note in Table 2.1 that the global average copper grade in granitoids is 0.64%, which is lower than the average value observed near the boundary. This is due to the relatively low grades in the eastern part of the study area far from the tourmaline breccia and porphyry. The decrease in grade in granitoids far from the boundary with the tourmaline breccia, on the one hand, and the closeness of the average grades observed near the boundary between the granitoids and tourmaline breccia (0.8%–0.9%), on the other hand, suggest a possible edge effect between these two facies. Therefore, a more careful analysis is needed, which will be based on two new tools: cross-correlograms and lagged scatter plots (definitions in Appendix, Section A.2.1) between the grades of different facies.

To determine the spatial cross-correlation between the copper grades observed in two different facies, three sets of totally heterotopic data are constructed (the grade in GD, the grade in TOB and the grade in POR), by eliminating the data located outside the facies of interest. Then, their cross-correlograms are calculated. It follows that the grades in the granitoids and the tourmaline breccia show significant correlations, but they have very low correlations with the grades in the porphyry (Figure 2.15).

The second tool is the lagged scatter plot of the grades, obtained by considering the pairs of data that have a small separation distance (less than 20 m) but belong to different facies. Again, the result indicates a similar average and a good correlation between the grades measured in the granitoids and the tourmaline breccia, whereas the grades of the porphyry facies are poorly correlated and have a much lower average value (Figure 2.16).

In summary, the copper grades in the porphyry have a much lower average and are only weakly correlated with the copper grades in the other two facies, which exhibit close average values and are strongly correlated. Of the tools discussed here, the ratio of the cross-variogram of the partial grade and facies indicator by the

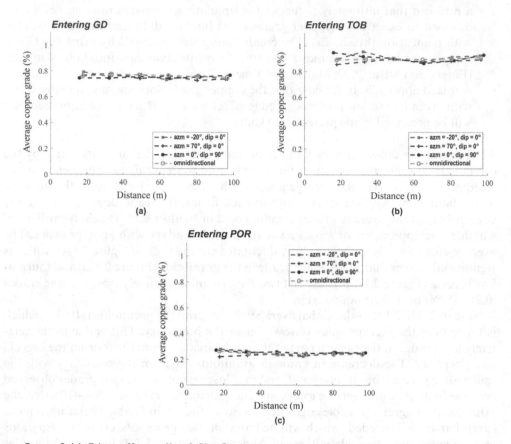

Figure 2.14 Edge effects (1 of 3). Cross-variogram of partial grade and facies indicator divided by indicator direct variogram, for (a) granitoids (GD),
(b) tourmaline breccia (TOB) and (c) porphyry (POR). Experimental
variograms are calculated along the N20°W, N70°E and vertical directions and omnidirectionally. For each facies, the average grade remains
constant when approaching the boundary, suggesting an absence of edge
effects.

indicator direct variogram provides information only on the average grade within
a facies. The cross-correlogram provides information only on the correlation between grades measured on each side of the boundary between two facies, whereas
the lagged scatter plot is more informative, as it shows both the averages and correlations between grades. In addition, for large separation distances, the cross-
correlograms and cross-to-direct variogram ratios are difficult to interpret because
their calculation involves pairs of samples at two points x and $x + h$ whose distances
to the facies boundaries may range from 0 to h. This is probably one of the reasons
why the edge effect between the tourmaline breccia and the granitoids is barely
noticeable in Figure 2.14.

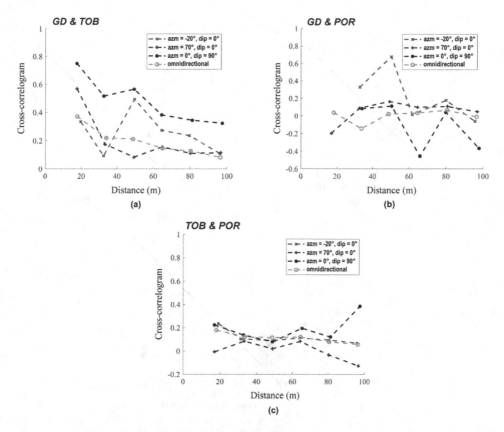

Figure 2.15 Edge effects (2 of 3). Cross-correlograms between the grades measured in different facies: granitoids (GD) and tourmaline breccia (TOB) (a), granitoids and porphyry (POR) (b), tourmaline breccia and porphyry (c). The correlograms are calculated along the N20°W, N70°E and vertical directions and omnidirectionally. There is a correlation between the grades measured in granitoids and those measured in the tourmaline breccia.

2.2.4 Grade modeling

The previous analyses lead us to a two-stage modeling: first, the grades in the porphyry facies are modeled, and then the grades in the other two facies, granitoids and tourmaline breccia, are modeled together.

2.2.4.1 First stage: modeling the grade in the porphyry

The copper grades in the porphyry are transformed into normal scores data (Gaussian anamorphosis, see Matheron, 1974, or Chilès and Delfiner, 2012), after ignoring all the data in the granitoids and the tourmaline breccia. The result is associated with a Gaussian random field Y_1' defined only in the porphyry. The direct variogram of Y_1' is then calculated and modeled, using a nugget effect and two nested spherical structures

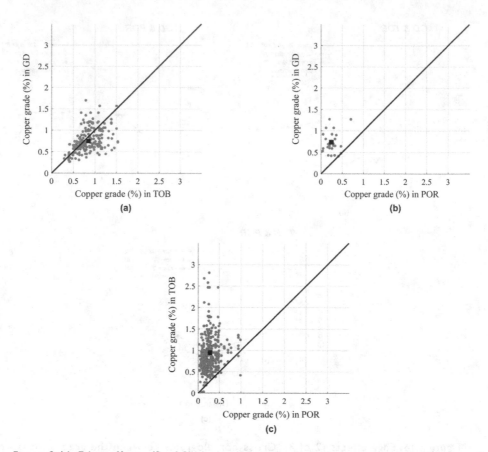

Figure 2.16 Edge effects (3 of 3). Lagged scatter plots, for a separation distance of
less than 20 m, between the grades measured in different facies: granitoids (GD) and tourmaline breccia (TOB) (a), granitoids and porphyry
(POR) (b), tourmaline breccia and porphyry (c). In each case, the black
square indicates the center of gravity of the scatter plot. The scatter plot
(a) shows correlated grades with a similar average value; the scatter plots
(b) and (c) show weak correlations and significant changes in the average
grades.

with a geometric anisotropy (Figure 2.17). As it is further assumed that the grade in the
porphyry is independent of the facies indicators, the cross-variograms between Y_1' and
Y_1 and between Y_1' and Y_2 are identically zero.

2.2.4.2 Second stage: modeling the grades in granitoids and tourmaline breccia

The copper grades measured in the granitoids and the tourmaline breccia are, in turn,
transformed into normal scores, now ignoring the porphyry data. The result is associated with a Gaussian random field Y_2' defined only in the granitoids and the tourmaline
breccia. The direct variogram of Y_2' and the cross-variogram of Y_2' and the tourmaline

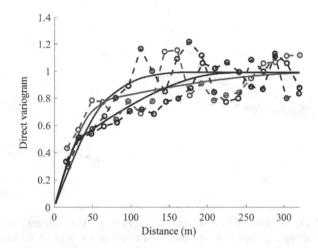

Figure 2.17 Fitting of the direct variogram of the normal score transform of the copper grade in the porphyry (Gaussian random field Y_1'). The experimental variograms, represented by dotted circles and lines, and modeled variograms, represented with solid lines, are calculated along the N20°W, N70°E and vertical directions, identified as the main directions of anisotropy.

breccia indicator $1_{Y_2<y_2}$ are calculated, ignoring the data in the porphyry. Taking into account the mathematical relationship between the cross-variogram of $1_{Y_2<y_2}$ and Y_2' and the cross-variogram of Y_2 and Y_2', a linear coregionalization model is fitted for both Gaussian random fields Y_2 and Y_2', based on the nested structures already used to fit the variogram of Y_2, in addition to a nugget effect (Figure 2.18).

Notably, in the fitted model, the spherical nested structure with the smallest range has no contribution to the cross-variogram between Y_2 and Y_2' and the correlation between both random fields (therefore, between the tourmaline breccia indicator and copper grade) stems from the spherical nested structure with the largest range. In other words, the dependence between copper grades and facies is associated with large scales of spatial variation.

To complete the coregionalization model of (Y_1, Y_2, Y_1', Y_2'), it is assumed that the Gaussian random field Y_2' is independent of the porphyry indicator and of the grade in the porphyry, which implies that the cross-variograms of Y_1 and Y_2' and of Y_1' and Y_2' are identically zero.

2.2.5 Joint simulation of facies and grades

It becomes now possible to simulate the facies and copper grades conditionally on the drill-hole data. For each desired simulation, the steps are as follows:

1. Simulation of the porphyry facies: The Gaussian random field Y_1 is simulated, conditionally on the porphyry indicator data (Table 2.2), and then truncated at threshold y_1 (Figure 2.12).

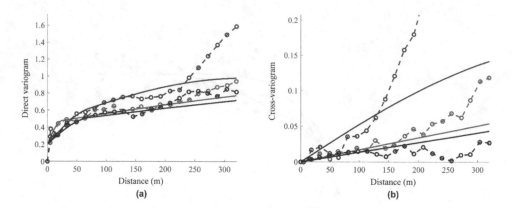

Figure 2.18 (a) Fitting of the direct variogram of the normal score transform of copper grade in granitoids and tourmaline breccia (Gaussian random field Y_2'). (b) Fitting of the cross-variogram of Y_2' and the tourmaline breccia indicator. The experimental (circles and dotted lines) and modeled (solid lines) variograms are calculated along the N20°W, N70°E and vertical directions.

2. Simulation of the copper grade in the porphyry facies: The Gaussian random field Y_1' is simulated, conditionally on the known values of this random field in the porphyry, and backtransformed into copper grade.

3. Cosimulation of the other two facies and their grades: In the space not occupied by the porphyry, the Gaussian random fields Y_2 and Y_2' are jointly simulated, conditionally on the known values of Y_2' and the indicator of Y_2 (with three states: 0, 1 or undefined, see Table 2.2). The simulation of Y_2 is transformed into facies (tourmaline breccia or granitoids) by truncation at the threshold y_2, whereas the simulation of Y_2' is backtransformed into copper grade.

The simulations of the copper grade so obtained show an edge effect when passing from the tourmaline breccia to the granitoids, and a lack of edge effect when entering the porphyry, marked by a 'hard' transition to very low grades (Figure 2.19). These characteristics are consistent with the chronology of the facies, since the porphyry is a postmineralization intrusive.

Averaging the values of the facies indicators or of the simulated copper grades at each point of the grid allows evaluating the probabilities of occurrence of the facies and their expected copper grades. The probability maps (Figure 2.20a–c) suggest relatively well-defined boundaries between facies, as they rapidly evolve from very low probabilities to very high probabilities, which is explained by the amount of available data and the strong spatial continuity of the facies indicators (Figure 2.13). In contrast, the copper grades are more erratic, with a significant nugget effect (Figure 2.18), which causes a strong smoothing effect when the simulations are averaged (Figure 2.20d).

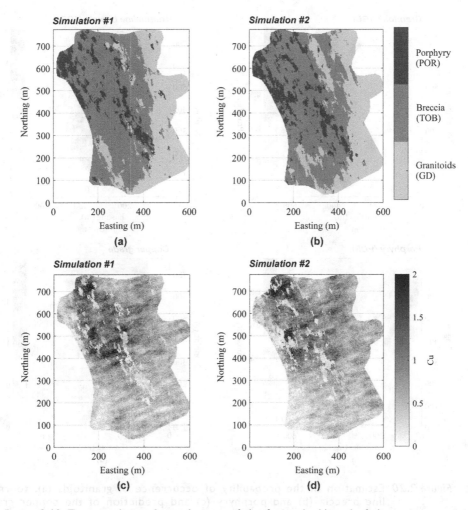

Figure 2.19 Two conditional simulations of the facies (a, b) and of the point-support copper grade (c, d), on a grid with a mesh of $2 \times 2 \times 16$ m^3 restricted to the sampled domain. The simulated grades reproduce the edge effect at the boundary between granitoids and tourmaline breccia, and the absence of edge effect at the boundary of the porphyry facies. These boundaries vary from one simulation to another. Horizontal projection of the bench with elevation 3,650 m.

2.2.6 Simulation validation

To validate the results, the so-called split-sample technique (also known as jack-knife) is used. The data set is partitioned into two subsets called the 'training set' and 'validation set', respectively. The partition is carried out by grouping the data of each drill hole and assigning, at random, each drill hole to one of the two subsets, with 1/3 and 2/3 probabilities, respectively. Then, the training set (1,023 data) is used as conditioning information to simulate the copper grades at the points of the validation set

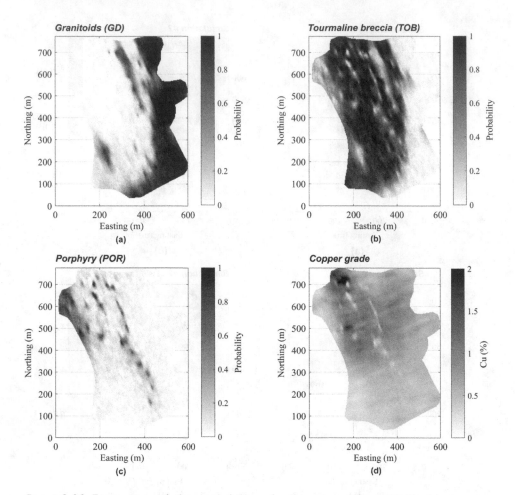

Figure 2.20 Estimation of the probability of occurrence of granitoids (a), tourmaline breccia (b) and porphyry (c) and prediction of the copper grade
(d), calculated from 50 simulations. The probability maps suggest fairly
well-defined facies, whereas the map of predicted grades is smooth due
to the greater variability of the true grades at small distances. Horizontal
projection of the bench with elevation 3,650 m.

(2,062 data), following the steps indicated earlier. In total, 500 simulations are built.
Two tests are carried out with the results. The first one aims at verifying the absence of
bias and conditional bias of the simulations, whereas the second one aims at verifying
that the fluctuations between the different simulations correctly measure the uncertainty of the copper grades at the points of the validation set.

2.2.6.1 First test

The copper grade can be predicted at the points of the validation set by averaging the
500 simulations, which constitutes an approximation to the conditional expectation,

a predictor without bias or conditional bias that minimizes the variance of the prediction error. This is corroborated by visualizing the scatter plot between the predicted grades and the true grades of the validation set: the regression curve of this scatter plot practically coincides with the diagonal line (Figure 2.21).

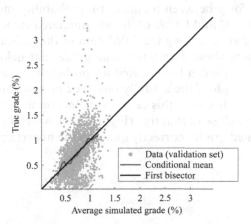

Figure 2.21   Scatter plot between the true copper grades (vertical axis) at the points of the validation set (2,062 data) and the average of 500 simulations (horizontal axis) conditioned on the data of the training set (1,023 data). The regression curve coincides with the diagonal (first bisector), which indicates the absence of conditional bias and, a fortiori, the absence of global bias.

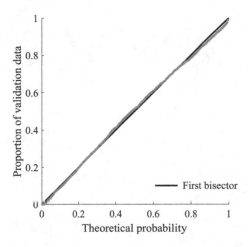

Figure 2.22   Proportion of data of the validation set belonging to a p-probability interval (vertical axis) as a function of the theoretical probability p (horizontal axis), for p varying between 0 and 1. The experimental points are located very close to the diagonal.

2.2.6.2 Second test

It is important to verify not only that the simulated grades fluctuate without global or conditional bias around true grades but also that their fluctuations correctly reflect the uncertainty associated with true grades. This second point can be verified by constructing, at each point of the validation set, probability intervals for the copper grade. Specifically, for p between 0 and 1, a p-probability interval bounded by the quantiles of order $1 - p/2$ and $1 + p/2$ of the 500 simulated values is constructed. From a statistical point of view, among the 2,062 data of the validation set, it is expected that a proportion p of these data has a true grade that belongs to the interval so constructed. The exercise can be repeated for probabilities ranging from 0 to 1 and summarized in the graph of the actually observed proportion of data as a function of the nominal probability p. In this case, the points on the graph are located on the diagonal line, or very close to that line (Figure 2.22), which indicates that the fluctuations of the simulated grades correctly quantify the uncertainty associated with the true grades.

Chapter 3

Grades and recoverable resources

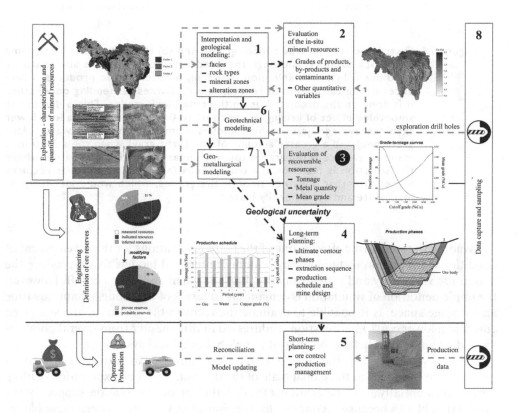

The grade predicted by kriging or cokriging suffers from a well-known smoothing effect. In particular, although kriging reproduces the average grade and leads to the smallest squared errors, the resulting block model does not reproduce the actual dispersion of the grades in the deposit. One consequence is the emergence of biases when applying a cutoff grade, a threshold value that separates the ore from the waste. Selecting the ore based on the kriging predictions tends to overestimate the true recoverable ore tonnage when the cutoff is lower than the overall average grade, or underestimate it when the cutoff is higher than the average grade (Figure 3.1).

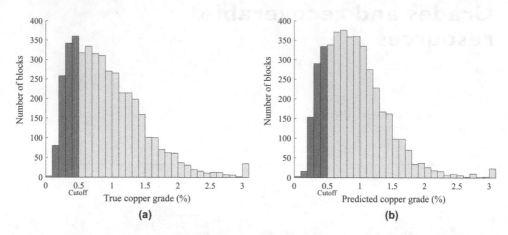

Figure 3.1 Distributions of (a) true copper grades and (b) their predictions by kriging at a block support. Although the predictor does not have any global or conditional bias, the selection based on a cutoff grade produces differences in the calculation of recoverable resources, depending on whether it is done on the predicted or on the true block grades. This is due to the smoothing effect of kriging. Here equal to 0.5%, the cutoff grade is lower than the average grade, which implies a smaller amount of waste (dark gray zone) and a greater amount of ore (light gray) when the selection is made on the predicted grades (b), in comparison with the situation where the selection is made on the true grades (a). Avoiding this bias requires a predictor that does not smooth the grades, which would inevitably be inaccurate and conditionally biased.

Avoiding the smoothing effect is one of the main motivations for the development of conditional simulation techniques, by building numerical models that reproduce the real grade variability and provide unbiased results when applying a cutoff. However, the implementation of simulation is demanding in terms of modeling, computing time and storage space. Is it possible to evaluate the recoverable resources above a given cutoff at a lower cost? This question is addressed in this chapter from several points of view, depending on whether the interest is a global or a local assessment.

The first section defines the concept of recovery functions and presents how to estimate these functions at the global scale of the deposit, based on exploration drilling data. It is essentially about determining the distribution of grades at the support of the production blocks because accounting for the change of support between these blocks and the drill-hole samples is crucial.

The second section focuses on predicting the recoverable resources locally, for each block of the deposit, and presents the multi-Gaussian kriging technique. Here, a challenge is to develop a predictor that is applicable when the random field associated with the grades is not strictly stationary, but only locally stationary.

A synthesis of the proposed approaches is presented in the third section.

3.1 Global recoverable resources

3.1.1 Data and problem setting

We remain in the Río Blanco-Los Bronces porphyry copper deposit in central Chile. This time, the case study focuses on another open-pit mine of the Codelco-Andina Division, called *Sur-Sur*. The data consist of around 2,400 copper grade measurements in exploration drill-hole samples distributed in a region of dimensions $400 \times 600 \times 130 \text{ m}^3$ (Figure 3.2). The sample support has been regularized to a length of 12 m along the drill holes, which corresponds to the bench height in this portion of the open pit. The high grades are mainly concentrated in the center of the region, corresponding to the tourmaline breccia, while the periphery contains lower grades and corresponds to granitoids, magmatic and monolithic breccias; the porphyry facies is absent here.

It is of interest to predict the recoverable resources above a given cutoff grade, the threshold value below which the ore extraction ceases to be economically profitable. At the scale of the sampled region, the problem amounts to estimating the distribution of the grades, first at the support of the drill-hole samples and then at the support of the production blocks.

3.1.2 First problem: inferring the distribution of the measured grades

According to the constitutive model of geostatistics, the grade $z(x)$ measured at a point x in space is interpreted as a realization of a random variable $Z(x)$. The measurement point actually corresponds to the volume of a drilling core sample small enough to be considered a 'point' support. The uppercase letter (Z) denotes a random variable, as opposed to the numerical value that it may take (z, lowercase). The set of random

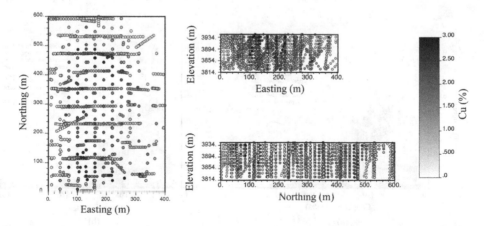

Figure 3.2 Data from the *Sur-Sur* mine of Codelco-Andina. Maps of the drill-hole samples in a local coordinate system: projections in horizontal and vertical sections. The scale indicates the copper grade.

variables associated with all the points of the studied region constitutes a random field, which here is assumed strictly stationary so as to simplify the inference of its parameters (see Appendix, Section A.2.7). In particular, the distribution of $Z(x)$ does not depend on the position of point x, and this is the distribution that has to be estimated from the available data. Among the experimental representations of this distribution, the most classical is the histogram (Figure 3.3a), which suggests here a lognormal distribution with an average of 1.05% and a standard deviation of 0.64%.

Is this histogram an accurate representation of the desired point-support grade distribution? The reply is negative due to the irregular sampling design, a frequent issue in the evaluation of mineral deposits: the high-graded sectors in the center of the region are more densely sampled due to their economic interest. Accordingly, the histogram in Figure 3.3a underrepresents the low grades of the periphery and gives a biased image of the true grade distribution in the region. To reduce this error, the experimental histogram is recalculated by weighting each sample, with a greater weight when the sample is isolated and a smaller one when it is clustered with other samples that essentially carry redundant information (Chilès and Delfiner, 2012). Figure 3.3b shows that the tail of the experimental distribution is reduced, resulting in a decrease of the average and the standard deviation to 0.90% and 0.59%, respectively.

The previous calculation still poses a practical problem. The stationarity hypothesis implies that a single distribution is enough to describe the random variable $Z(x)$ at any point x. But the same does not happen with the true grade distribution because the periphery of the sampled region has lower grades than the center; increasing the size of the region would therefore induce a significant decrease in its average grade and modify the grade distribution. The geostatistical modeling hypotheses, in particular the stationarity of the random field associated with the grade, shows here their limits.

Accordingly, the following modeling and calculation of the 'global' recoverable resources refer to a region that contains the drill-hole data and whose exact contour is not explicitly defined. The objective of the global calculations is to guide the decision

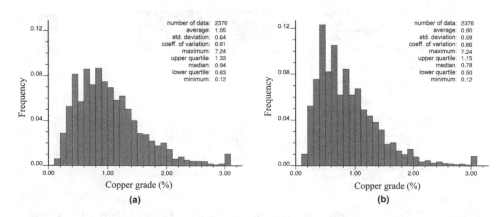

Figure 3.3 Histograms of sampled copper grades. (a) All the samples receive the same weight. (b) Isolated samples are weighted more than clustered samples; the weighting is inversely proportional to the number of samples contained in a $40 \times 40 \times 20$ m^3 block centered on the target sample.

of the engineers about the viability of mining the deposit, but these calculations are intended to be later complemented by 'local' predictions.

3.1.3 Recovery functions

Hereinafter, attention is paid to the 'recovery functions', also called the 'transfer functions'. Associated with a cutoff z (lowercase), they describe the grade distribution. Four of them are of particular interest (Matheron, 1984; Lantuéjoul, 1990; Chilès and Delfiner, 2012):

- the *fraction of tonnage* above z, which represents the fraction of ore whose grade exceeds the cutoff grade or, equivalently, the probability that, at a given point x, $Z(x)$ is greater than z:

$$T(z) = E\left\{1_{Z(x)>z}\right\} \tag{3.1}$$

where $1_{Z(x)>z}$ is the indicator equal to 1 if $_{Z(x)>z}$, 0 otherwise;
- the *metal quantity* above z:

$$Q(z) = E\left\{Z(x)\, 1_{Z(x)>z}\right\} \tag{3.2}$$

- the *average grade* above z:

$$m(z) = \frac{Q(z)}{T(z)} = E\{Z(x)\,|\,Z(x) > z\} \tag{3.3}$$

- the *conventional benefit* above z:

$$B(z) = Q(z) - zT(z). \tag{3.4}$$

The functions T, Q, m and B do not depend on the chosen point x due to the stationarity hypothesis (even if, as mentioned earlier, this hypothesis is questionable). For a zero cutoff grade, one has $T(0)=1$ and $Q(0)=m(0)=B(0)=m$ (global average grade).

3.1.4 Second problem: change of support

The question that interests the geologist and the mining engineer is how the distribution of grades and the recovery functions change, when passing from a quasi-point support (drill-hole sample) to a block support (volume V of a selective mining unit). In other words, the interest here is the estimation of the recovery functions at a block support, based on point-support information:

- fraction of tonnage above z:

$$T_V(z) = E\left\{1_{Z(V)>z}\right\} \tag{3.5}$$

- metal quantity above z:

$$Q_V(z) = E\left\{Z(V)\, 1_{Z(V)>z}\right\} \tag{3.6}$$

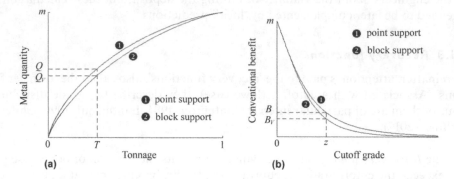

Figure 3.4 The selectivity of distributions: (a) hierarchy of the metal–tonnage curves according to support; (b) hierarchy of the conventional benefit curves according to the support.

- mean grade above z:

$$m_V(z) = Q_V(z) / T_V(z) = E\{Z(V) | Z(V) > z\} \tag{3.7}$$

- conventional benefit above z:

$$B_V(z) = Q_V(z) - zT_V(z). \tag{3.8}$$

Matheron (1984) and Lantuéjoul (1990) showed that the distribution of block grades is 'less selective' than that of the point-support grades. For the same extracted tonnage T, the quantity of metal Q recovered is less if the extraction is performed on the block support than if it is performed on a point support: $Q_V(T) \leq Q(T)$ and $Q_V(1) = Q(1) = m$ (Figure 3.4a). Similarly, for a given cutoff grade, the conventional benefit decreases when the support increases: $B_V(z) \leq B(z)$ and $B_V(0) = B(0) = m$ (Figure 3.4b). This geostatistical concept of selectivity formalizes what the geologists and mining engineers understand by 'dilution' (a mixture of high- and low-grade materials that deteriorates the selection above the cutoff grade).

3.1.5 Looking for a change-of-support model

Several simple models were designed in the 1970s to derive the block-support grade distribution from the point-support one. The most widely known are the affine and lognormal corrections (Journel and Huijbregts, 1978). Here, a (little) more sophisticated model is selected: the 'discrete Gaussian model', tested many times with success in the past decades. The starting point is the transformation of the grade random field (Z) at the point support (x) into a random field Y with a standard Gaussian distribution (Figure 3.5):

$$Z(x) = \phi(Y(x)), \tag{3.9}$$

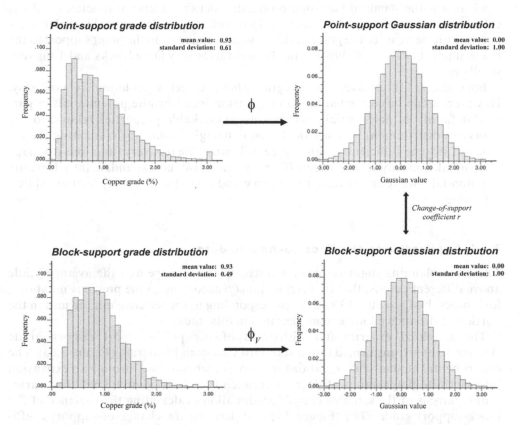

Figure 3.5 Explanatory scheme of the discrete Gaussian model. The distributions of point-support (top left) and block-support (bottom left) grades are transformed into standard Gaussian distributions (right) through the point- and block-support anamorphosis functions. The change-of-support coefficient *r* is used to relate the Gaussian variables defined on the two supports.

where ϕ is a nondecreasing function known as the Gaussian transformation function or 'anamorphosis' (Matheron, 1974). A similar transformation is done for the block-support grade:

$$Z(V) = \phi_V(Y_V), \tag{3.10}$$

where Y_V is a random field with a standard Gaussian distribution, and ϕ_V is the block-support anamorphosis function, the knowledge of which allows retrieving the block-support grade distribution and the desired block-support recovery functions. Matheron established in 1976 the following identity between the point- and block-support anamorphoses:

$$\phi_V(y) = \int \phi\left(r\,y + \sqrt{1 - r^2}\,t\right) g(t)\mathrm{d}t, \tag{3.11}$$

where g is the standard Gaussian probability density (a known function) and r is called the change-of-support coefficient. This coefficient is determined in order to reproduce the decrease of the grade variance when passing from the point support to the block support (Rivoirard, 1994); it lies between 0 for very large blocks and 1 for very small ones.

Formula (3.11) involves an integral whose direct calculation is not easy. However, it can be converted into an expansion into Hermite polynomials (a particular family of polynomials that possess remarkable properties related to the Gaussian distribution), whose calculation is straightforward by truncating the expansion at some sufficiently high degree. Likewise, the recovery functions (tonnage and metal quantity above a cutoff grade) can also be expanded into Hermite polynomials; we refer the reader to Emery and Soto-Torres (2005) for the explicit formulae.

3.1.6 Application to the Codelco-Andina data

The mine planning engineers are interested in the tonnage and the average grade above different cutoffs, that is, $T_V(z)$ and $m_V(z)$ according to the previous notations, for a block V of size $10 \times 10 \times 12$ m^3 corresponding to the selective mining unit in the portion of the open pit under consideration in this study.

The weighted experimental histogram of the point-support copper grade (Figure 3.3b) is transformed into a standard Gaussian histogram (Figure 3.6a). The experimental anamorphosis, a stair function, is then modeled through an expansion into Hermite polynomials of degrees between 0 and 50 (Figure 3.6b). The variogram analysis of the sampled copper grades allows calculating the variance of the block-support grade $Z(V)$ (Figure 3.7) and deriving the change-of-support coefficient associated with the block V of size $10 \times 10 \times 12$ m^3: one finds $r=0.86$. This value, close to 1, indicates that the block is small in relation to the correlation range of the grades.

The evaluation of the ore tonnage and of the metal quantity above different cutoff grades is then carried out (Figure 3.8). As expected, the resulting distribution of the block-support grades is less dispersed than that of the point-support grades, with a higher tonnage at low cutoffs and a lower tonnage at high cutoffs. Quantifying this phenomenon is interesting and helps to figure out the importance of the 'dilution' produced by mining blocks instead of point supports. For instance, 74.8% of the drill-hole samples exceed a cutoff grade of 0.5%, with an average copper grade of 1.06%, which provides a point-support metal quantity of 0.793%. In contrast, for the block support, 81.3% of the grades exceed the same cutoff, with an average of 1.01%, giving a metal quantity of 0.821%. The above metal quantities are expressed in the same unit as the copper grade (percent), in accordance with the definitions (3.2) and (3.6) prized by geostatisticians: one has to multiply them by the total rock tonnage of the area to be mined and to divide by one hundred in order to figure out the total copper tonnage (in tonnes) that can be recovered when selecting the ore above the cutoff.

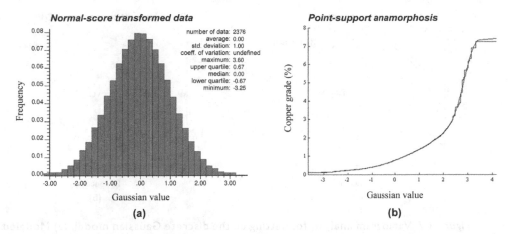

Figure 3.6 Gaussian transformation of the sampled copper grades. (a) Grade histogram after transformation, (b) point-support anamorphosis function. The experimental anamorphosis is a stair function (broken lines) and is modeled by an expansion into Hermite polynomials of degrees from 0 to 50 (continuous curve). The original grade histogram is shown in Figure 3.3b.

3.2 Local recoverable resources

The estimation of the recovery functions at the scale of the deposit only provides a general description of the recoverable resources, without details of the sectors that may contribute to the tonnage or recovered metal. It is therefore important to be able to complete the global study through a local study where, for each block of the deposit, the expected recovery functions for this block are evaluated. These would correspond to the average of the recovery functions obtained over a large number of simulations of the block grade conditioned to the grades measured at the sampling data points. Here again, one wants to avoid the use of simulations and find an analytical solution to locally predict the recoverable resources. Is this challenge hopeless, to paraphrase a paper by Rossi and Parker (1994)?

3.2.1 One easy solution: multi-Gaussian kriging

Two problems overlap: estimating the local grade distributions conditionally on the available sampling data, and incorporating the change of support between the point-support data and the blocks to be mined. The literature offers several techniques capable of simultaneously addressing both problems, in particular if one deals with a disseminated deposit for which the multi-Gaussian random field model is suitable (Rivoirard, 1994; Chilès and Delfiner, 2012; Rivoirard et al., 2014). In the case of the Río Blanco-Los Bronces deposit and the Codelco-Andina data, this assumption is validated by examining the experimental bivariate distributions (lagged scatter plots) calculated on pairs of Gaussian data separated by 20 m (Figure 3.9a) and 100 m (Figure 3.9b): they exhibit the typical (elliptical) shape of the bivariate Gaussian

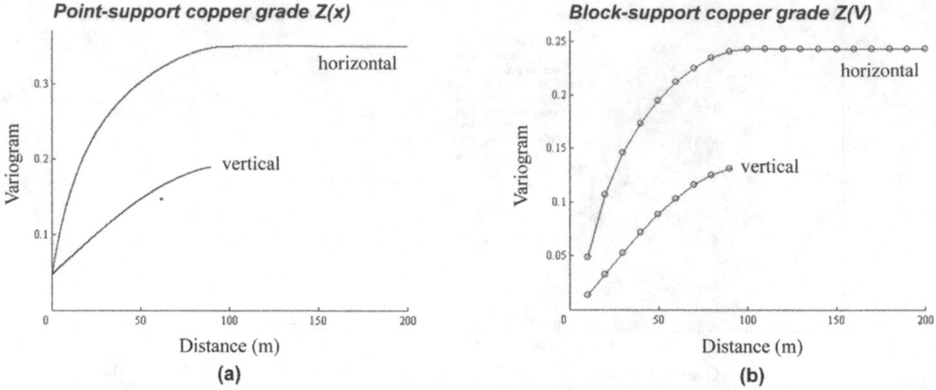

Figure 3.7 Variogram analysis for setting up the discrete Gaussian model. (a) Modeled variogram of the point-support grades. (b) Numerical calculation of the grade variogram regularized on blocks of $10 \times 10 \times 12$ m^3. Only the sill of this regularized variogram is needed at this stage.

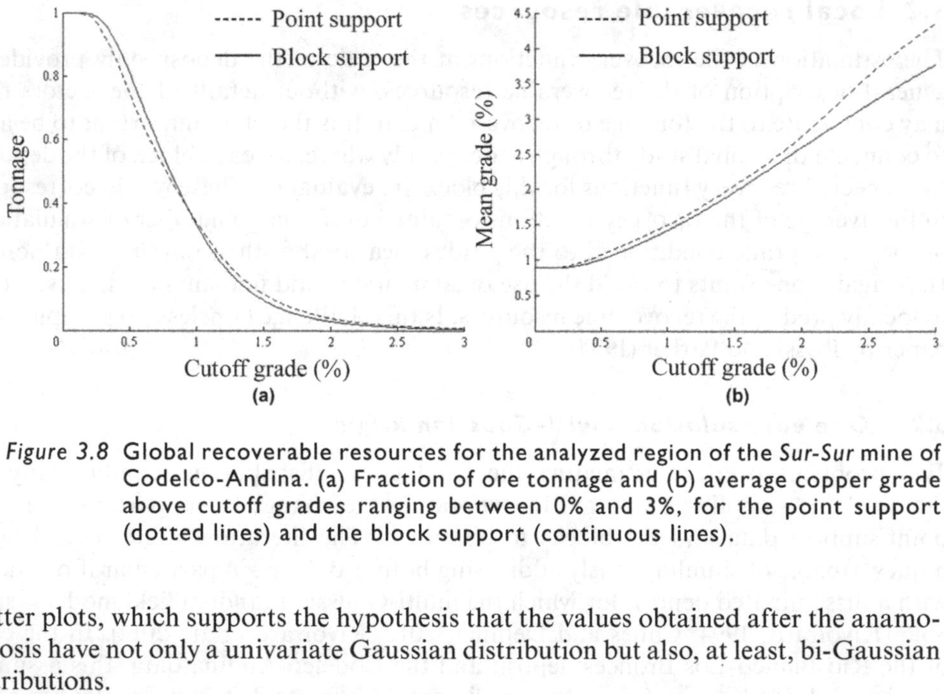

Figure 3.8 Global recoverable resources for the analyzed region of the *Sur-Sur* mine of Codelco-Andina. (a) Fraction of ore tonnage and (b) average copper grade above cutoff grades ranging between 0% and 3%, for the point support (dotted lines) and the block support (continuous lines).

scatter plots, which supports the hypothesis that the values obtained after the anamorphosis have not only a univariate Gaussian distribution but also, at least, bi-Gaussian distributions.

In the following, two specific recovery functions are considered: the tonnage (T) and the metal quantity (Q) above cutoff z. T and Q are the expected values of the indicator $1_{Z(V)>z}$ and the grade $Z(V)$ multiplied by the indicator $1_{Z(V)>z}$, respectively, which are functions of the Gaussian transform Y_V and can be written in the general form $\varphi(Y_V)$.

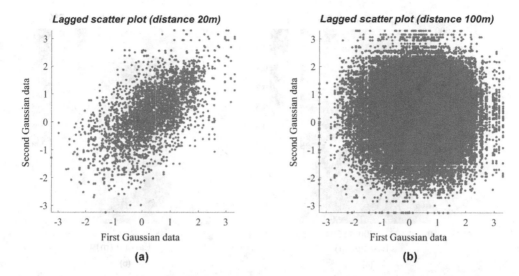

Figure 3.9 Binormality of the normal scores of the sampled copper grades. Lagged scatter plots for a separation distance of (a) 20 m and (b) 100 m (omnidirectional calculations, with a tolerance of ±2.5 m in each case).

The determination of the recoverable resources of a specific block V therefore amounts to predicting a function $\varphi(Y_V)$ by using the point-support data in and around V.

The discrete Gaussian model presented in the previous section can be generalized if the random fields Y_V and $Y(\underline{x})$ (with a point support $\underline{x}$ randomized uniformly in V) are jointly multi-Gaussian. It is then possible to calculate the expectation of $\varphi(Y_V)$ for any block V, conditionally on the available data on $Y(\underline{x})$:

$$\left[\varphi(Y_V)\right]^* = \int \varphi\left(Y_V^{SK} + \sigma_V^{SK}\, t\right) g(t)\mathrm{d}t, \tag{3.12}$$

where Y_V^{SK} and σ_V^{SK} stand for the simple kriging of Y_V from the data on $Y(\underline{x})$ and for the standard deviation of the simple kriging error, respectively. Interestingly, this predictor (3.12), known in the literature as the 'conditional expectation' (Rivoirard, 1994) or as 'multi-Gaussian kriging' (Verly, 1983; David, 1988), is often absent in geostatistical software. One reason invoked by a developer of such software is the difficulty of programming the integral (3.12). However, we realized that this integral, like the one in (3.11), can be rewritten as an expansion into Hermite polynomials, which allows its straightforward calculation. The reader is referred to Rivoirard (1994) and Emery (2005b) for details on how to perform the variogram analysis required for kriging and on the formulae to calculate the recoverable tonnage and metal by multi-Gaussian kriging.

As an illustration, Figure 3.10 maps the results associated with a cutoff of 0.5%, for the blocks located on the bench with elevation 3,922 m. These maps cannot be obtained by applying the cutoff to the kriged copper grade, which would deliver biased predictions, as mentioned due to the smoothing effect of kriging.

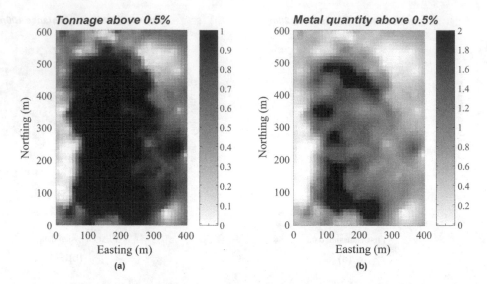

Figure 3.10 Recoverable resources for the analyzed region of the *Sur-Sur* mine of Codelco-Andina (1 of 2). Multi-Gaussian kriging of (a) the ore tonnage and (b) the metal quantity above a 0.5% cutoff grade, for blocks of 10 × 10 × 12 m^3 located on the bench of elevation 3,922 m.

3.2.2 Weakening the stationarity assumption

The central part of the region, where the high grades are found, has the highest values of tonnage and metal quantity above the cutoff. However, some occurrences of high tonnages and metal quantities also appear at the edge of the region, particularly in the northeast corner where the drill-hole data are scarce. In this corner, the multi-Gaussian kriging, based on a simple kriging of Y_V that converges to the average 0 when one moves away from the data, provides results close to the global tonnage and metal quantity above the 0.5% cutoff grade. These global values are high because the average copper grade in the region is about 0.9%, more than the cutoff. Again, it is the stationarity hypothesis that is questioned here, according to which the Gaussian random fields at the point and block supports have a zero mean in the entire space. Can this hypothesis, too stringent, be weakened?

At the time of the study (2002), a solution existed when the function φ to be evaluated was an exponential, a special case known as 'lognormal kriging': one can then assume that the mean is constant only locally, at the scale of the kriging neighborhood, but varies at the scale of the entire region, and substitute the simple kriging Y_V^{SK} with ordinary kriging Y_V^{OK} and the simple kriging variance $\left(\sigma_V^{SK}\right)^2$ with the ordinary kriging variance $\left(\sigma_V^{OK}\right)^2$ plus twice the Lagrange multiplier μ_V^{OK} of the ordinary kriging system, which still provides an unbiased predictor of $\exp(Y_V)$

(Rivoirard, 1994). Curiously, no point in the proof of this disruptive result uses the particular properties of the exponential function involved... This is how we discovered (Emery, 2005b) that an unbiased predictor of $\varphi(Y_V)$ for any function φ can be constructed by setting:

$$\left[\varphi(Y_V)\right]^{**} = \int \varphi\left(Y_V^{OK} + \sqrt{\left(\sigma_V^{OK}\right)^2 + 2\mu_V^{OK}}\ t\right) g(t)\mathrm{d}t \tag{3.13}$$

The application of such an 'ordinary multi-Gaussian kriging' to the Codelco-Andina data shows an improvement of the predictions in the edge of the region (Figure 3.11): the ore tonnages and metal quantities obtained are now lower, which is consistent with the geological intuition, as the mineralization is concentrated in the center of the region. The undersampled edge in the east corresponds to granodioritic rocks of low copper grade. The maps in Figure 3.11 provide the geologists and mining engineers with a quantification of the recoverable resources (ore tonnage and metal tonnage) on a block-by-block basis, accounting for the cutoff grade used to distinguish ore from waste, while the smooth block model constructed with block kriging would overestimate these tonnages. As for the assessment of the global resources, it is necessary to multiply the calculated ore and metal tonnages (expressed here in percent) by the tonnage of a block (approximately 3,000 tonnes if one considers a rock density of 2.5 t/m^3) and to divide by one hundred, in order to get these tonnages expressed in tonnes.

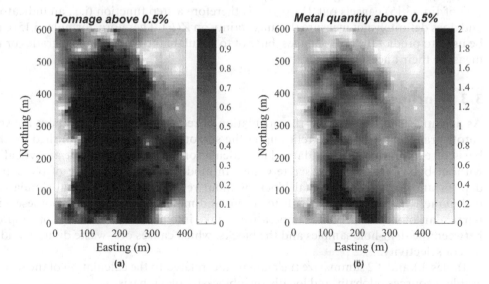

Figure 3.11 Recoverable resources for the analyzed region of the *Sur-Sur* mine of Codelco-Andina (2 of 2). Ordinary multi-Gaussian kriging of (a) the ore tonnage and (b) the metal quantity above a 0.5% cutoff grade, for blocks of 10 × 10 × 12 m^3 located on the bench of elevation 3,922 m.

3.2.3 Discussion: prediction and quantification of uncertainty

The tonnage of a block V above a cutoff grade z is obtained by predicting the indicator $1_{Z(V)>z}$. By varying the cutoff on the set of positive real numbers, an estimate of the complementary cumulative distribution function of the block grade is obtained. In this case, the multi-Gaussian kriging expressions (3.12) and (3.13) simplify as follows:

$$\left[1_{Z(V)>z}\right]^* = G\left(\frac{Y_V^{SK} - y}{\sigma_V^{SK}}\right)$$

(3.14)

$$\left[1_{Z(V)>z}\right]^{**} = G\left(\frac{Y_V^{OK} - y}{\sqrt{\left(\sigma_V^{OK}\right)^2 + 2\mu_V^{OK}}}\right)$$

(3.15)

where G is the Gaussian cumulative distribution function and y is the transform of z by the block-support anamorphosis ϕ_V defined in Eq. (3.10).

Viewed as two functions of z, are they also able to measure the uncertainty in the true block grade? The answer is positive in the case of $[1_{Z(V)>z}]^*$, but negative for $[1_{Z(V)>z}]^{**}$. In particular, if a single data is available in the kriging neighborhood, the ordinary kriging predictor is equal to this data and the ordinary kriging variance is equal, up to the sign, to twice the Lagrange multiplier, so that the denominator in the right-hand side of Eq. (3.15) cancels out: $[1_{Z(V)>z}]^{**}$ is therefore a step function (i.e., an indicator) and does definitely not measure the uncertainty on $Z(V)$. Therefore, formula (3.15) can be used to predict the ore tonnages, but not to simulate the block-support grades or to measure their uncertainty.

3.3 Synthesis

As a synthesis of this chapter, the calculation of recoverable resources above a given cutoff grade is biased if one relies upon the smooth block model established on the basis of exploration drilling data. This calculation can be carried out analytically, without bias, by using the discrete Gaussian model, both at the scale of the entire deposit and for each individual block, without resorting to conditional simulation techniques that are more difficult to set up and more demanding in hypotheses, in particular, of stationarity. It is imperative to take into account the change of support between the drill-hole samples and the blocks, which causes ore–waste dilution and a loss of selectivity.

Tables 3.1 and 3.2 summarize the main results related to the calculation of the recoverable resources, globally and locally on a block-by-block basis.

Table 3.1 Summary of the global estimate of recovery functions by the discrete Gaussian model

Objective	From point-support exploration data, estimate, at the scale of the entire deposit, the recovery functions of the production blocks
Preferred approach	Discrete Gaussian model
Restrictions of use	Stationarity of the univariate grade distribution, i.e., invariance of this distribution by a translation in space
Stages for implementation	1. Gaussian anamorphosis o Weight the data to compensate for irregularities in the spatial sampling design o Calculate the experimental point-support grade distribution o Transform it into a standard Gaussian distribution o Expand the anamorphosis into Hermite polynomials 2. Determination of the change-of-support coefficient o Calculate the (weighted) experimental variogram of the point-support grades o Fit a variogram model o Calculate the variance of the block-support grades by regularizing the modeled variogram o Derive the change-of-support coefficient 3. Calculation of global recovery functions: ore tonnage and metal quantity

Table 3.2 Synthesis of the local prediction of recoverable resources by the discrete Gaussian model

Objective	From the point-support exploration data, predict the fraction of ore tonnage $I_{Z(V)>z}$ and the metal quantity $Z(V)I_{Z(V)>z}$ for any block V of the deposit and any cutoff grade z
Preferred approach	Multi-Gaussian kriging
Restrictions of use	• Disseminated deposit, compatible with a diffusion (multi-Gaussian) random field model • Small blocks, for which the randomization of the sample positions in the blocks induces little loss of information • Local stationarity of the grade
Stages for implementation	1. Anamorphosis of the point-support grade and calculation of the change-of-support coefficient (Table 3.1) 2. Validation of the bi-Gaussian assumption for the point-support transformed data, for example, by examining their lagged scatter plots 3. Variogram analysis of Y_V 4. Multi-Gaussian kriging, preferably assuming that the mean values of the transformed grades are unknown to obtain a robust prediction against spatial variations of these means (Eq. 3.13)

Chapter 4

Long-term planning and reserves

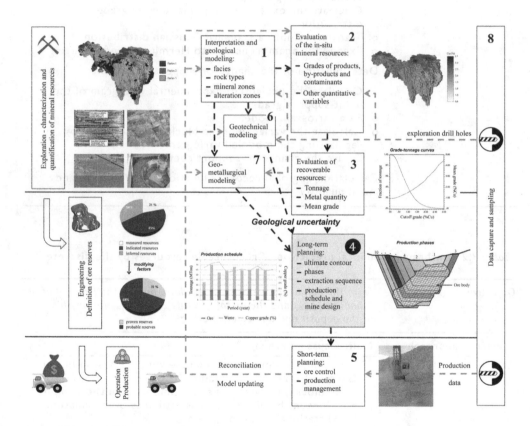

The block model is one of the main inputs of the long-term or strategic planning process. The objective is to define, for the life of the mine, which blocks should be extracted, when, and what should be their destinations (mineral concentration plant, heap, dump, stockpile, etc.), the usual objective being to maximize the net present value (NPV) of the mining project and to limit its risks.

Planning techniques depend on the mining method. In open-pit mining, they are carried out through the following stages (Hustrulid et al., 2013):

- Economic valuation of the blocks: The value of a block depends not only on the grades of the main products, by-products and contaminants it contains but also on the rock density, the metallurgical recovery (fraction of the metal content that can actually be recovered), the metal sale price, the sale costs, extraction costs and processing costs. The rock density can vary depending on the rock type or on the block grade (as it happens, in particular, in iron deposits), and the extraction and processing costs can be a function of the position of the block in the deposit and of its lithological and mineralogical properties.
- Determination of the ultimate or final pit of the open-pit mine, that is, of the geometric contour of the mine at the end of its lifetime (Figure 4.1a): This contour allows estimating the amount of extractable ore from the deposit, defining the economic duration of the exploitation, sizing the facilities and equipment necessary for production and planning the production capacity.
- Determination of exploitation phases that divide the final pit into successive intermediate pits, also known as pushbacks (Figure 4.1b): These nested pits can be obtained, for instance, by calculating the ultimate pits associated with increasingly lower prices of the produced metal(s) or with increasingly higher cutoff grades.
- Definition of the mining strategy, corresponding to the sequence according to which the blocks will be extracted, from the origin of the exploitation until the final pit is reached (Figure 4.1c).
- Definition of a production schedule, which details the quantities of extracted material (ore and waste), their destinations, ore grades and economic values (cash flows) for the entire life of the mine, in general, by annual periods.

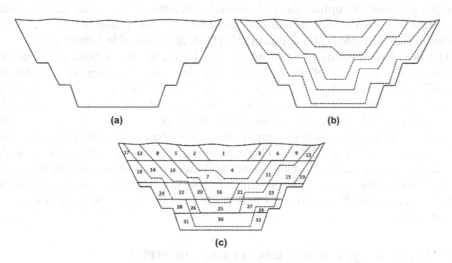

Figure 4.1 Planning an open-pit mine. Schematic vertical section: (a) final pit of the mine, (b) phases or pushbacks, (c) mining strategy or block extraction sequence. (Credit: Mohammad Maleki.)

A comment on vocabulary: the production schedule defines the 'ore reserves', that is, the extractable fraction of measured or indicated resources, taking into account numerous technical–economic criteria: accessibility of the resources, grades of the products and by-products, presence of contaminants, geological and metallurgical characteristics, sale prices of metals, and legal, social and environmental considerations, among others.

The long-term planning is a problem of optimization under restrictions: it aims to maximize the benefit (NPV of the mining project) subject to restrictions on the extraction of the blocks: for example, the maximum slope angle, the precedence relationships between the blocks due to their locations in the deposit, or the maximum capacity of the mineral concentration plant. The combinatorial of such a problem is considerable, especially since the deposit can be discretized into several millions or tens of millions of blocks, each of which is characterized by one to several tens of geological, geotechnical and geometallurgical variables. Therefore, in general, it is necessary to resort to numerical analysis techniques such as operational research techniques and graph theory.

In addition, the maximization of the benefit does not guarantee that this benefit will be achieved in reality, or even that the mining project will be successful, insofar as several of the parameters used to value the blocks are not known with certainty, in particular:

- the future values of the costs and sales price of the produced metal(s): this is the 'market uncertainty';
- the geological properties, including the grades and rock types, which will be referred to as the 'geological uncertainty' hereunder.

The question then arises to quantify the impact of such uncertainties on the mining plan. The plan should be not only optimal but also robust against happenstances or hazards. Concerning market uncertainty, a common practice is to use different cost and price scenarios – optimistic, intermediate, pessimistic. For geological uncertainty, scenarios can also be constructed by geostatistical simulation, which will be illustrated in the Ministro Hales deposit (MMH) already visited in Chapter 1.

The next two sections present, on the one hand, the geostatistical modeling and conditional simulation of the grades of five elements of interest in a sector of the Ministro Hales deposit and, on the other hand, the foundations of long-term planning and its application to determine a final pit and a production schedule based on a reference block model, namely the average of the grade simulations. The results are combined to study the impact that geological uncertainty may have on the production schedule so defined. The penultimate section presents some thoughts to optimize the final contour of the mine, no longer on the basis of a reference block model, but on the basis of a set of simulations that reflect the geological uncertainty. The last section addresses the problem of the classification of mineral resources and ore reserves.

4.1 Modeling geological uncertainty in MMH

The exploration drill-hole database contains over 50,000 samples composited at a length of 1.5 m. The available information corresponds to the grades of one main product (copper), two by-products (molybdenum and silver) and two contaminants (arsenic and antimony), as well to the rock types and mineral zones (Tables 4.1 and 4.2).

The following study (Montoya et al., 2012) is limited to a sector of the sampled area of approximately $700 \times 250 \times 600$ m^3 (Figure 4.2). The statistics of the grades of the selected samples (Table 4.3) show the presence of very high silver, molybdenum, arsenic and antimony grades. Furthermore, the correlation between copper and

Table 4.1 Ministro Hales deposit: rock types

Code	Rock type
BXC	Central breccia
BXCS	Southern central breccia
BXCW	Western central breccia
CI	$0.1 < Cu < 0.49$
C5	$0.50 < Cu < 2.50$
MY	Myriam breccia
PMM	Mansa Mina porphyry

Table 4.2 Ministro Hales deposit: mineral zones

Code	Mineral zone
AAS	Sulfide ore with high arsenic grade
BAS	Sulfide ore with low arsenic grade
OXI	Oxide ore
MIX	Mixed ore

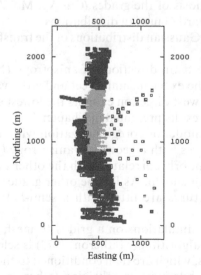

Figure 4.2 Ministro Hales deposit. Sampled area (dark gray) and sector of study (light gray). Horizontal projection.

Table 4.3 Ministro Hales deposit: statistics of measured grades (sector under study)

Variable	Number of samples	Minimum	Maximum	Average	Variance
Cu (%)	26,004	0.00	30.34	1.24	3.26
Ag (g/t)	25,022	0.01	2,143.6	27.21	2,558.2
Mo (g/t)	23,214	0.92	4,894.4	55.58	14,216.5
As (g/t)	25,311	4.60	1.16×10^5	1,158.4	1.01×10^7
Sb (g/t)	10,724	0.92	7,580.8	103.88	1.00×10^5

silver grades is 0.73, the correlation between copper and arsenic grades is 0.68, and the correlation between copper and antimony grades is 0.61. These correlations can be explained by the presence of minerals whose chemical formula contains copper, silver, arsenic and/or antimony, in particular: enargite (Cu_3AsS_4), luzonite (Cu_3AsS_4), tennantite ($(Cu, Fe)_{12}As_4S_{13}$), argentotennantite ($(Ag, Cu)_{10}(Zn, Fe)_2(As, Sb)_4S_{13}$) and famatinite ($Cu_3SbS_4$).

Most of the samples are located in the sulfide mineral zone. Moreover, the distribution of rock types (facies) turns out to be very irregular, with numerous small-scale alternations. A hierarchical approach where the grades would be simulated in each facies separately, for example, through a residual indicator model (Chapter 2, Eq. 2.13), is discarded in order to avoid dividing the deposit into innumerable objects of little thickness. Consequently, the geostatistical modeling is carried out according to the following stages:

- Regularization of the sample length to 6 m.
- Gaussian anamorphosis of the grades (Cu, Ag, Mo, As, Sb), providing five new variables with standard Gaussian distributions.
- Validation of the bi-Gaussian distributions of the transformed data (elliptic lagged scatter plots).
- Identification of the main directions of anisotropy (N0°E, N90°E and vertical) and calculation of the experimental direct and cross-variograms of the Gaussian variables: The east–west direction exhibits the lowest spatial continuity, which is consistent with the results presented in Chapter 1.
- Fitting of a linear model of coregionalization with a nugget effect and nested exponential structures with a geometric anisotropy (Figure 4.3): Since the molybdenum grade is poorly correlated with the other grades, its fitting is realized separately and is not shown. As for the other grades (Cu, Ag, As, Sb), the sills of the nested structures are fitted with a semi-automatic algorithm (Emery, 2010).
- Construction of 40 simulations on a grid with mesh $2 \times 6 \times 6$ m^3 (Figure 4.4): The turning bands algorithm (Matheron, 1973) is selected to build the nonconditional simulations, which are then conditioned to the drill-hole data by kriging (in the case of molybdenum) or cokriging (other grades) (Chilès and Delfiner, 2012).

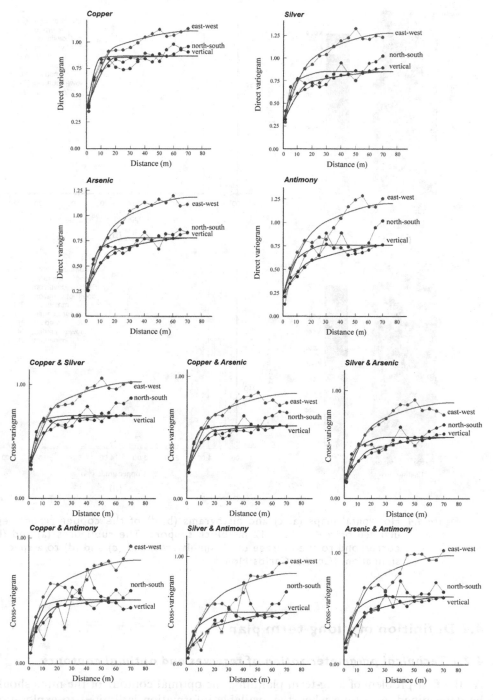

Figure 4.3 Experimental (points) and modeled (continuous lines) direct and cross-variograms of the Gaussian variables along the main directions of anisotropy (N0°E, N90°E and vertical).

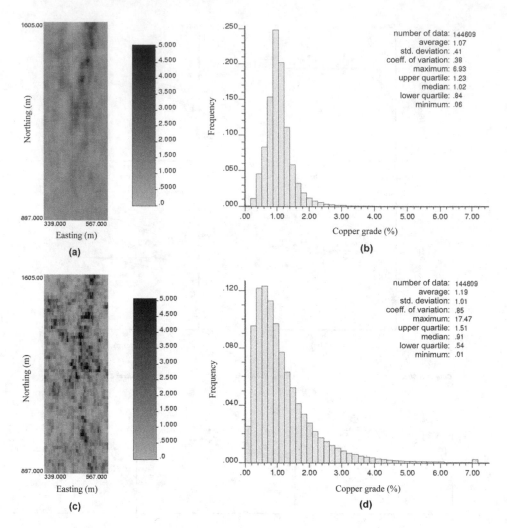

Figure 4.4 Horizontal maps (a, c) and histograms (b, d) of the copper grades regularized on a 4 × 12 × 12 m³ block support. The subfigures (a) and (b) correspond to the average of 40 simulations, and (c) and (d) to a specific simulation. (Credit: Carlos Montoya.)

4.2 Definition of a long-term plan

4.2.1 Technical parameterization of reserves and optimal contours

As the first problem of long-term planning, the optimal contour of the mine should be determined, at a time when the available information is limited to exploration drilling data. A solution to this problem, called 'technical parameterization of the reserves' by Georges Matheron (1975a–c) and further implemented by Dominique

François-Bongarçon (1978) and Thierry Coléou (1987), consists in obtaining this best final contour in two different stages:

1. Determination of a family of optimal contours depending on a limited number of technical parameters: The final contour that is sought necessarily belongs to this family of technically optimal contours, irrespective of the formula used for the economic valuation of resources.
2. Choice of the economic parameters and definition of the best final contour, which will be the contour that maximizes the chosen valorization formula, among all the previously identified technically optimal contours.

Let C be the set of possible contours for the mine, for example, the set of all the pits that can be excavated in the case of an open-pit operation. For any contour $A \in C$, three essential technical parameters are considered:

*   the total extracted tonnage $T_t(A)$, which includes both the ore and the waste;
*   the tonnage of extracted ore $T_o(A)$;
*   the metal quantity recovered from the ore $Q(A)$.

It is also assumed that, for given values of the total tonnage t_t and ore tonnage t_o such that $t_t \geq t_o \geq 0$, the technically optimal contour is the one that maximizes the recovered metal quantity, that is:

$$Q^*(t_t, t_o) = \sup_{A \in C} \left\{ Q(A) : T_t(A) = t_t, T_o(A) = t_o \right\}. \qquad (4.1)$$

The optimum Q^* is a nondecreasing function of t_t and t_o, which vanishes at $(t_t, t_o) = (0,0)$ and is maximum when t_o is the total tonnage of the blocks with nonzero grades (Figure 4.5). The question arises of determining Q^* without comparing all the possible contours, which would be prohibitive in terms of computing time.

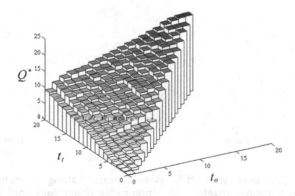

Figure 4.5 Representation of technically optimal contours in the space spanned by (Q, t_t, t_o).

The idea proposed by Matheron is to replace Q^* with its concave envelope Q^{**}, defined as the lower limit of all the planes tangent to the surface $Q^*(t_t, t_o)$:

$$Q^{**}(t_t, t_o) = \inf_{\lambda, \theta \geq 0} \left\{ \gamma(\lambda, \theta) + \lambda t_t + \theta t_o \right\} \tag{4.2}$$

with

$$\forall \lambda, \phi \geq 0, \gamma(\lambda, \theta) = \sup_{A \in C} \left\{ Q(A) - \lambda T_t(A) - \theta T_o(A) \right\}. \tag{4.3}$$

While Q^* is a stair function, Q^{**} is a piecewise linear function (Figure 4.6). The problem of the technical parameterization of reserves amounts to determining the surface $Q^{**}(t_t, t_o)$ (Eq. 4.2) or the dual surface $\gamma(\lambda, \theta)$ (Eq. 4.3).

A reformulation of Eqs. (4.2) and (4.3) allows removing the term θ from the problem of optimizing the contour of the mine. Indeed, if $m(A)$ denotes the average grade of the ore contained in contour A, then

$$Q(A) - \lambda T_t(A) - \theta T_o(A) = T_o(A)(m(A) - \theta) - \lambda T_t(A) = Q_\theta(A) - \lambda T_t(A) \tag{4.4}$$

where $Q_\theta(A) = T_o(A)(m(A) - \theta)$ is the amount of recovered metal associated with the grades decreased by θ. The problem of the technical parameterization of the reserves then reduces to minimizing the function of a single parameter λ:

$$\gamma(\lambda) = \sup_{A \in C} \left\{ Q_\theta(A) - \lambda T_t(A) \right\}. \tag{4.5}$$

In the case of an open-pit operation, the extraction of a block V requires previously extracting the blocks located in an area called 'extraction cone' of V and denoted $\Gamma(V)$

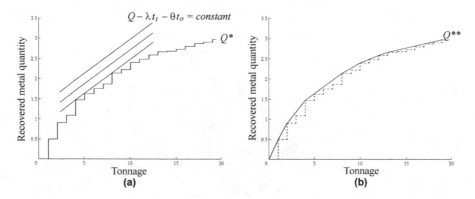

Figure 4.6 (a) Representation of Q^* (stair function) along the tonnage (t_t or t_o) axis. (b) Concave envelope Q^{**} (piecewise linear function) that substitutes Q^* in the determination of the technically optimal contours of the mining operation.

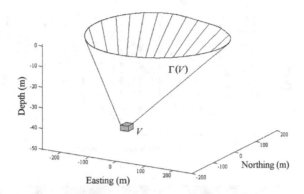

Figure 4.7 Extraction cone of a block *V*, representing the set of blocks to be mined before extracting *V*. The shape of the base and the angle of the cone depend on the geomechanical properties of the rock mass.

(Figure 4.7). It should be kept in mind that the higher the value of λ, the less deep the pit to excavate, which corresponds to more unfavorable economic conditions. Therefore, each block *V* of the deposit can be assigned a limit value $\Lambda(V)$ such that the block belongs to the optimum pit if and only if $\lambda \leq \Lambda(V)$. The optimum pit associated with a given parameter is then defined by:

$$C_\lambda = \{V : \lambda \leq \Lambda(V)\}. \tag{4.6}$$

The function Λ thus defined has interesting mathematical properties. On the one hand, a block *W* belonging to the extraction cone of *V* necessarily belongs to all the pits that contain *V*, so that $\Lambda(V) \leq \Lambda(W)$: the function Λ is Γ-increasing, that is, increasing according to the order relation induced by the definition of the cone Γ. On the other hand, the function Λ is constant in the zone between two successive optimum pits and equal to the average grade of this zone (Figure 4.8). These properties imply that the function Λ is the Γ-increasing function that is the 'closest' to the grade function. The determination of Γ boils down to a problem of projection of the grade function in the space of the Γ-increasing functions.

To solve this projection problem, several approaches based on maximum closure or maximum flow algorithms in a graph, dynamic programming or integer programming, were developed since the 1960s (e.g., Lerchs and Grossmann, 1965; Matheron, 1975a, b; François-Bongarçon, 1978; Wright, 1989; Underwood and Tolwinski, 1998; Hochbaum and Chen, 2000). By repeating the algorithm for increasing values of the λ parameter, a series of nested pits is obtained that can guide the definition of the phases and the exploitation strategy to reach the final pit and plan the open-pit mining (Caccetta, 2007). However, these nested pits do not necessarily constitute operational or technically realizable phases, in particular because they ignore certain geometric conditions for the displacement of the mining equipment, or because a slight variation of the parameter λ could contribute a considerable volume to mine (Bai et al., 2018).

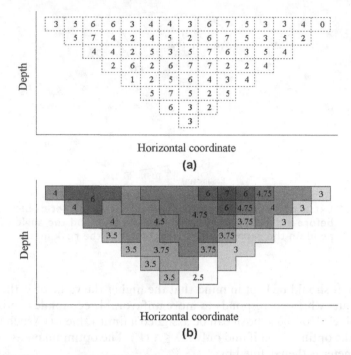

Figure 4.8 (a) Block model with their grades or valuations and (b) associated Λ function, equal to the average grade of each gray zone located between two optimal pits. Representations of a vertical section with extraction cones with a slope angle of 45°.

4.2.2 Application to MMH: final pit and production schedule

The definition of the final contour of the open-pit mine is based here on a 'reference block model', corresponding to the average of the 40 grade simulations constructed in the previous section of this chapter; this average is similar to a cokriging of the grades. The algorithm by Lerchs and Grossmann (1965) is used with the technical and economic parameters indicated in Table 4.4. The blocks are valued by a combination of their copper, silver and molybdenum grades. Once the final pit is obtained, a ten-year production schedule with four phases of comparable volumes is defined (Figure 4.9). The NPV of the production schedule amounts to US$1,204,660,000.

4.3 Sensitivity of the plan to geological uncertainty

We now address the questions that motivate this chapter: what impact can geological uncertainty have on the production schedule? Is it sure that the production promises and the NPV of the long-term plan will be fulfilled? Is this plan robust against uncertainty in the true grades contained in the deposit?

Table 4.4 Technical and economic parameters for the determination of the final pit of
the mine

Parameter	Value	Unit
Rock density	2.6	(ton/m^3)
Maximum slope angle	50.0	(degrees)
Copper price	1.3	(US$/lb)
Silver price	7.0	(US$/oz)
Molybdenum price	15.0	(US$/lb)
Copper metallurgical recovery	85.0	(%)
Silver metallurgical recovery	70.0	(%)
Molybdenum metallurgical recovery	50.0	(%)
Mining costs	1.1	(US$/ton)
Processing costs	6.0	(US$/ton)
Copper smelting and refinery cost	0.4	(US$/lb)
Silver smelting and refinery cost	0.4	(US$/oz)
Molybdenum smelting and refinery cost	2.0	(US$/lb)
Mine investment	1,500	(US$/extracted tpd)
Processing plant investment	4,500	(US$/ treated tpd)
Discount rate	10.0	(%)

The values of these parameters have been adjusted to preserve the confidentiality of the real
project, without undermining its plausibility.

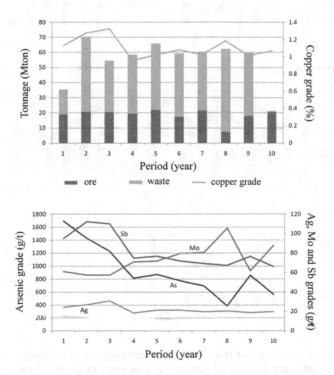

Figure 4.9 Production schedule defined from the reference block model. The gray
bars indicate the tonnages of ore and waste extracted during the ten years
of production, whereas the broken lines represent the average annual
grades of copper (a) and other elements (b). (Credit: Carlos Montoya.)

4.3.1 Uncertainty in the grades of the production schedule

A sensitivity analysis is carried out by applying the production schedule to each of the previously constructed grade simulations. In each case, the tonnages of ore and waste are the same, but the grade changes depending on the simulation under consideration. These changes are summarized in the following curves, obtained for 10 out of the 40 initially produced simulations (Figure 4.10). Although the reference block model on the basis of which the long-term plan is defined smoothes the grades, it is found that it accurately predicts (without bias) the grades that will be extracted. This property is explained by the absence of conditional bias of the reference block model (Montoya et al., 2012).

In addition, the simulations allow quantifying, for each annual period, the potential deviations between the grades of the ore to be extracted and the grades provided by the

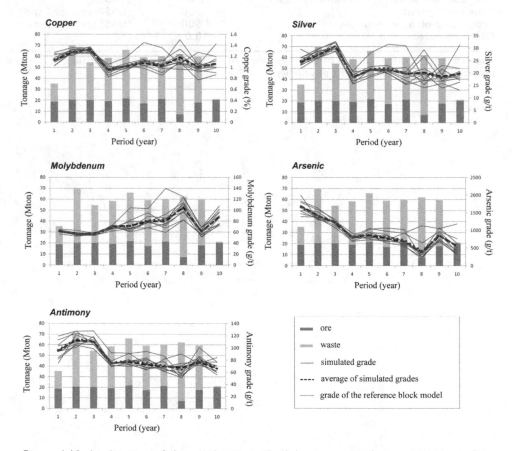

Figure 4.10 Application of the production schedule to ten simulations interpreted as as many geological scenarios. The average grade of each simulation (black dotted curve, average of the ten continuous gray curves) coincides with the grade obtained from the reference block model (gray dotted curve). The plan constructed from this block model therefore provides unbiased predictions of the grades that will be effectively extracted.

production schedule, as well as estimating the probability that the actually extracted grades are greater or less than the planned grades in a certain period. For example, for the first period (year 1), three of the ten simulations exceed the expected value of the arsenic and antimony grades, indicating a probability of about 30% of not achieving the proposed objective for these two grades.

Too high arsenic or antimony grades have an impact on the quality of the concentrate obtained by flotation of the ore and in the smelter, where a fraction of the arsenic and antimony is emitted into the atmosphere. This is the reason why most copper smelters apply severe restrictions to the arsenic grades of the concentrates that they accept to process. The previous sensitivity analysis is then useful for planning engineers: it is not enough for them that the restrictions in the contaminant grades (As and Sb) are fulfilled by the reference block model; it is also necessary that they be fulfilled by most of the simulations in order to minimize risks.

4.3.2 Uncertainty in the net present value of the mining plan

The simulations also allow determining the financial risk of the production schedule. The NPV calculated with ten simulations coincides, on average, with the value calculated using the reference block model, which is again explained by the absence of conditional bias of this model and by the fact that, here, the NPV is a piecewise linear function of the grades (Montoya et al., 2012). Variations, positive or negative, of the NPV obtained in the simulations can reach up to 20% of the value predicted by the reference block model (Table 4.5). This difference, 20%, corresponds to the order of magnitude of the expected errors made during the prefeasibility and feasibility studies, which are aimed at demonstrating the technical and economic feasibility of a mining project, while inaccuracies of the order of 30%–50% may occur at the prior conceptual study stage that aims to provide a preliminary description of the project.

Table 4.5 Uncertainty in the net present value of the production schedule

	Net present value (MUS$)	Variation with respect to the reference block model (%)
Reference block model	1,204.66	0.00
Simulation n°1	1,111.80	−8.35
Simulation n°9	1,173.17	−2.68
Simulation n°10	1,247.11	3.40
Simulation n°19	1,186.56	−1.53
Simulation n°27	1,028.60	−17.12
Simulation n°31	1,507.65	20.10
Simulation n°33	1,326.84	9.21
Simulation n°35	1,269.20	5.09
Simulation n°37	1,340.93	10.16
Simulation n°39	1,034.75	−16.42
Average	1,222.67	1.47

The value calculated in a simulation may differ from the value calculated in the reference block model by up to MUS$300, representing 20% of the reference value. However, on average over ten simulations, the calculated value is very close to the value of the reference model, which therefore delivers an unbiased prediction of the true value

4.3.3 Discussion

The previous results highlight the importance of the absence of conditional bias of the block model used in long-term planning, a property that can be used to predict without bias the grades to be extracted and the NPV of the mining project. In general, the absence of conditional bias is emphasized for short-term planning, when deciding the destination of the extracted blocks, but it is rarely mentioned as a significant factor for long-term planning, which, however, is an important property. It should be remembered that, often, conditional bias is caused by a poor design of the moving neighborhood used in kriging or cokriging (see Section A.3 in Appendix).

4.4 Looking for the optimal final contour

In the previous section, a mining plan was sensitized to different geological scenarios to quantify the changes in tonnages, average grades and NPVs that can occur between one scenario and another. But the long-term plan has been defined from a smoothed block model and has no reason to be the optimal plan for each scenario.

The search for an 'optimal' plan that incorporates geological uncertainty in open-pit mining is not trivial, since the algorithms for determining the final pit (Eq. 4.6) and the successive intermediate pits are formulated for a unique block model containing the information of the block grades.

Of course, it is always possible to apply these algorithms to several geological scenarios obtained by conditional simulation. The final pits and production schedules are compared based on waste tonnages, ore tonnages, ore grades and NPVs. An example for the Ministro Hales deposit is shown in Table 4.6. It is seen that the ore grades in the final pit are higher and the ore tonnage is smaller when the final pit is optimized in each simulation instead of the reference block model that corresponds to the simulation average. These differences, which have a significant impact on the NPV of the mining project, are explained by the smoothing effect caused by averaging simulations: the occurrence of intermediate grades increases (high and low grades become scarcer), which causes an increase in the tonnage above the cutoff grade that defines the ore and a decrease in the average grade of this ore.

Hence, the final pit that the planning engineer usually calculates from a smoothed block model is far from being optimal. In the previous example, the NPV obtained with the reference block model (Million US$ 1,204.66) can overestimate or underestimate

Table 4.6 Results of the optimization of the final pit, for the reference block model (average of 40 simulations) and for two specific simulations

	Reference block model	Simulation n°31	Simulation n°39
Total extracted tonnage (Mton)	547.16	562.82	358.16
Ore tonnage (Mton)	181.68	157.84	121.30
Copper grade (%)	1.11	1.42	1.25
Silver grade (g/t)	22.66	30.47	26.19
Molybdenum grade (g/t)	71.73	74.40	70.41
Arsenic grade (g/t)	973.37	1,167.51	1,191.14
Antimony grade (g/t)	83.62	102.81	100.87
Net present value (MUS$)	1,204.66	1,507.65	1,034.75

by up to 20% the value of a production schedule developed on the basis of a single simulation. This percentage represents the order of magnitude of the financial uncertainty
of the mining project due to uncertainty in the grades.

An adaptation of a well-known algorithm to determine the final mining pit, the so-
called 'floating cone' method (Carlson et al., 1966), which allows the optimization of
the pit over a set of geological scenarios, is presented in the following. This adaptation
was originally proposed by Reyes et al. (2012) and Reyes (2017).

The floating cone is an iterative algorithm based on the geometric shape of the extraction cone of a block (Figure 4.7). At each iteration, a new cone is placed in a block
chosen at random in the block model: if this cone improves the economic value of
the pit, it is accepted and the pit is increased with this cone; in the opposite case, the
cone is rejected and the pit is not modified. A known drawback of the algorithm is
that it does not necessarily converge toward the optimal pit, a drawback that can be
eliminated using the so-called 'simulated annealing' technique. Let us consider the
following ingredients:

- an initial state: a pit equal to the empty set;
- a transition kernel, which allows proposing a new pit from the current pit, consisting in choosing at random between removing one of the cones of the pit or adding
  a new cone positioned randomly in the block model;
- an objective function O that, for any possible pit, measures its 'value';
- a positive scalar $t(0)$ known as the initial 'temperature';
- a 'cooling schedule', that is, a decreasing function that indicates, for iteration k,
  the temperature $t(k)$ used for this iteration, for example, $t(k) = t(0) \, \varepsilon^k$ with $\varepsilon < 1$;
- a stopping criterion that can be, for instance, the maximum number of iterations.

At iteration k, the transition that improves the objective function is always accepted,
as in the original floating cone algorithm. However, a transition that deteriorates the
objective function is not always rejected: it is accepted with probability $\exp(-\Delta O/t(k))$
where ΔO represents the deterioration of the objective function (loss of value of the pit).

This algorithm allows two other notable improvements for the definition of the pit.
The first one is the incorporation of geological uncertainty through the definition of
the objective function. In fact, this function allows not only the valuation of a pit for
a single block model but also its valuation for a set of geological scenarios obtained
by simulation. This could be, for example, the value of the pit obtained on average in
all the scenarios, or the worst value if the planning engineer has a strong aversion to
risk. In summary, the choice of the objective function can incorporate the planner's
preferences or the policy of the mining company.

The second improvement refers to the operationalization of the final pit. Defining a
pit as the union of extraction cones as shown in Figure 4.7 is only an approximation to
the problem of designing an open-pit operation. Spiky cones provide a global geomechanical stability of the pit, but not its operational capacity: a minimum basic contour at
the bottom of the mine should be considered to operate the excavators and the vehicles
that transport the extracted material, as well as a minimum bench height and protection
and containment berms to ensure slope stability. Instead of spiky cones, it is possible to
apply the floating cone algorithm with more realistic objects (pseudocones) that include
these geometric characteristics (Figure 4.11). Likewise, the application of mathematical

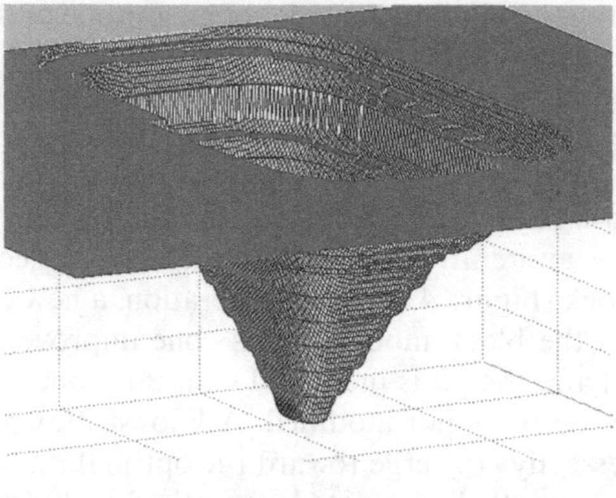

Figure 4.11 Pseudocone including a minimum base (pseudoellipse similar to a rectangle with rounded angles), a ramp and benches with walls of fixed height and with safety berms. (Credit: Manuel Reyes.)

morphology tools (dilation, erosion, opening and closing) can be useful to eliminate protuberances or cavities with acute angles and to smooth the contour of the pit obtained (Reyes, 2017; Bai et al., 2018), so that it is not only optimal but also operational.

The previous example is intended to provide an overview of the potential of geostatistical tools, in particular, of conditional simulation, when combined with optimization techniques to define the final contour of the mine, the extraction sequence and the production schedule.

For more than one decade, an important research effort in 'stochastic mine planning' has been carried out, where the long-term plan is no longer defined on the basis of a single smooth block model, but on the basis of many geological scenarios that represent geological uncertainty (Dimitrakopoulos, 2007, 2011, 2018; Dimitrakopoulos and Ramazan, 2008; Espinoza et al., 2013a, b; Marcotte and Caron, 2013; Vargas et al., 2014; Aguirre et al., 2015; Moreno et al., 2017; Maleki et al., 2020).

4.5 Classification of mineral resources and ore reserves

Planning in the long term also means converting mineral resources into ore reserves. The technical and public reports on the prediction of resources and reserves should consider several categories according to the degree of confidence in the prediction.

For the mineral resources, three categories are defined:

- 'measured' resources, predicted with a significant confidence;
- 'indicated' resources, predicted with a reasonable confidence;
- 'inferred' resources, predicted with a low confidence.

The ore reserves are divided into two classes:

- 'proven' reserves, originating from measured resources;
- 'probable' reserves, originating from indicated resources, but sometimes also from measured resources.

The definition of the categories is a very subjective exercise, with greater reason because the international codes giving the classification guidelines do not provide an exact definition of the concepts of significant or reasonable confidence, leaving the definition in the hands of an expert called a 'competent' or a 'qualified' person, a statute that is acquired by affiliation with a professional association or the cooptation of a peer, as well as by the justification of an adequate university degree and a regular and consistent industrial mining practice.

It is worth noting that nothing in these codes indicates how to make the predictions, or sometimes in a very relaxed way, which gives a considerable amount of latitude to these calculations and places results obtained in very different ways at the same level. In the absence of being able to propose a universal classification criterion, we here limit ourselves to a few thoughts:

- Geostatistics provides a set of tools and techniques to measure the uncertainty in the grade of each block of the deposit. Some practitioners use the variance of the kriging error, whereas others prefer the variance obtained on a set of conditional simulations, the conditional coefficient of variation (square root of the conditional variance divided by the average of the simulated grades), the width of a given probability interval or the width divided by the average value, to name a few examples.
- The classification is sensitive to the chosen uncertainty measure and to the threshold values used to separate the categories (Emery et al., 2006). In particular, there is no one-to-one relationship between the above measures.
- The classification is also sensitive to the support of the mining block (David, 1988). The uncertainty on the grades can be reduced for a large block but becomes high when this block is subdivided into smaller ones, especially if the spatial grade correlation is low. In practice, the uncertainty measure is often established for a block support considered relevant for the technical and economic evaluation, for example, a quarterly, semiannual or annual volume of production.
- The quality of the sampling, from the geological mapping of the drill-hole cores to the chemical analysis of the grades, the degree of knowledge of the geology of the deposit and the quality of the interpreted geological model must be taken into account when classifying the mineral resources. In the case of the ore reserves, one should also consider the degree of knowledge of the modifying factors, including the economic factors and the factors related to the mining, mineral and metallurgical processes (Müller and de Nordenflycht, 2006).

Chapter 5

Short-term mine planning

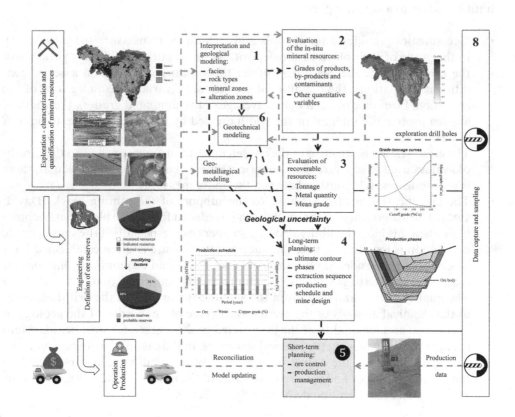

Regardless of whether the extracted ore is processed by flotation (sulfide ore) or by heap leaching (oxide ore), it is important, during production, that essential properties such as the grades, lithology or mineralogy be kept as close as possible to those for which the metal recovery is maximum. The 'ore control' aims at classifying the extracted blocks according to their geological and geometallurgical properties, with the objective of guaranteeing the economic grade of the ore sent for treatment, while minimizing the fluctuations of these properties. It may be necessary to mix ores from several blocks before sending them to treatment, which implies managing stockpiles with well-defined characteristics.

In open-pit mining, ore control is carried out on the basis of information from 'blast holes', in addition to the exploration drill holes. Already evoked in Chapters 2 and 3, the blast holes are realized on a fine mesh, of the order of 5–10 m, in order to introduce the explosives that will fragment rock for extraction (Figure 5.1). The ore control operation includes an analysis of the material extracted from the hole, in particular, an assessment of its grade and/or its mineralogy.

The following are two examples of the use of blast-hole data for short-term planning and production management. The first concerns the joint modeling of the copper grades measured in drill-hole samples and blast holes and their use to update the block model. The second example is interested in the mineralogy of the ore sent to treatment (heap leaching) and in estimating the risk of excess of given minerals – copper clay and copper wad – that have a negative influence on the copper recovery and acid consumption.

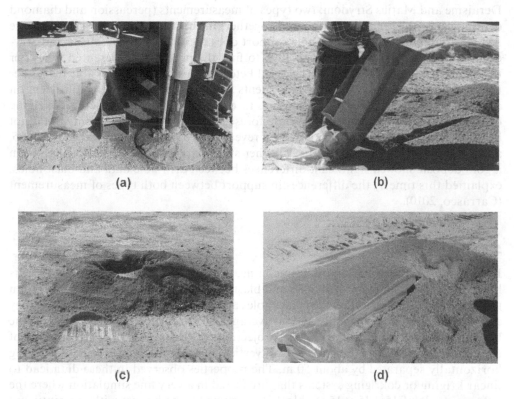

(a) (b)

(c) (d)

Figure 5.1 Blast holes. (a) Perforation and extraction of rock fragments using a radial tray. (b) Recovery of a sample (of the order of 20 kg) from the fragments contained in the radial tray. (c, d) Detritus of rock fragments (a few tonnes) around the hole into which an explosive charge will be introduced to fragment the rock. (Credit: Eduardo Magri.)

5.1 Updating the block model for production

In Chapter 2, the blast-hole grades were used to calculate an average value of each block considered as the reality and to decide the destination of the blocks to be mined, which is a generalized practice in the mining industry. The question here is to know if these measurements can really serve as a reference and how consistent they are with the measurements from the diamond drill holes, which are considered more reliable.

The two types of measurement already differ in the length of their cylindrical support: of the order of 15 m for blast holes, against 3 m for the drill-hole measurements in our example. The diameters of the blast holes and drill holes are small (a few centimeters to tens of centimeters) and play a secondary role compared with the length of the measurement. If the information on drill holes and blast holes is consistent, it should be possible to link it to their respective supports through regularization formulae, allowing combined use of both measurements to update the block model, initially built on the basis of drill-hole data, at the time of exploitation. The challenge of this study is important.

The literature on the subject is rather poor, except for an interesting work by Jacques Deraisme and Marius Strydom: two types of measurements (percussion and diamond drill holes) of the same grade are used together. In this case, the essential difference is due to sample quality, and sample support does not intervene (Deraisme and Strydom, 2009). One has to go back to 1995 to find the problem as presented here. For the Chuquicamata mine, Danie Krige and Peter Dunn studied the possible biases between drill-hole and blast-hole measurements and found that they were consistent on average, without involving their support. In conclusion, they proposed a cokriging of the two types of measurement, while foreseeing that the improvement would not be of primary importance, which will be revealed to be true (Krige and Dunn 1995, and Chapter 8). On the other hand, another study carried out by Pedro Carrasco in a gold deposit showed significant differences between drill-hole data and blast holes, explained this time by the differences in support between both types of measurement (Carrasco, 2010).

5.1.1 Data, method

Let us go back to the Radomiro Tomic mine, where an area has been selected for its homogeneous sampling by drill holes and blast holes. Four times more numerous than the drill-hole samples, the vertical blast holes are distributed in space differently: they are separated 10 m horizontally and an average of 15 m vertically, which explains the black horizontal bands in the vertical projections of Figure 5.2. In contrast, most of the drill holes are distributed in east–west vertical planes and sampled every 3 m, being horizontally separated by about 50 m. The properties observed in these data lead to linear kriging or cokriging systems that are tested in a very fine simulation where the average grade of $15 \times 15 \times 15$ m^3 blocks is assumed to be known with certainty and constitutes the 'reality' against which the predictions are compared.

This study is based on the equality between the mean grades of blast holes and drill holes. If this is not the case, the use of the sampling theory, which will be addressed in Chapter 8, is essential to reduce the bias to a reasonable level, an extraordinary but confidential work done by Francis Pitard for Codelco prior to this study.

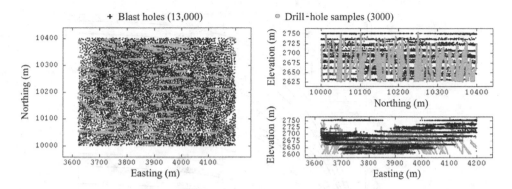

Figure 5.2 Projections of blast-hole and drill-hole measurements. Light gray: the drill-hole measurements (support 3 m); black: the blast-hole measurements (support 15 m).

5.1.2 Comparison

The verifications are based on the variogram, the cornerstone of the prediction, in a process that had its fashion in the 1970s (Matheron, 1971; Huijbregts, 1971; Journel and Huijbregts, 1978). Starting with the variogram of the measurement known on a first support (here, the 3 m drill-hole samples), the underlying point-support variogram is derived by theoretical deconvolution (i.e., regularization) and then transformed to the second support (15 m for blast holes) by convolution, to verify whether the theoretical result corresponds to the experimental variogram of the blast-hole data. Since these calculations are based on numerical approximations, the opposite can also be done (starting from the large support to reach the small support).

The calculations are tedious. Two cases must be distinguished, depending on whether the regularization is parallel or perpendicular to the direction of the vector that links the pairs of measurements involved in the calculations. Nomograms like the one in Figure 5.3 can be used. These nomograms are barely legible because no one has bothered to redraw the charts dating back to Jean Serra's work in 1967.

It was a surprise to discover, during this study, that Serra, the founder of mathematical morphology with Georges Matheron, wrote in his youth more than 500 pages on the subject, which can no longer be found in any of our institutional libraries. EBay offered an 'as new' copy, which was purchased for a small amount of money quickly sent to a retired Québec geologist.

Figure 5.4 shows the results. In both cases (vector parallel or perpendicular to the direction of the drilling), the shape of the curves is good, but the theoretical nugget effect (0.01) associated with the support of 15 m is underestimated. It should be increased to 0.03 for the theoretical and experimental variograms to match. When the calculations are carried out starting from the blast-hole data, the theoretical nugget effect obtained is not much more credible (Séguret, 2015).

The conclusion is that one part of the nugget effect measured on both supports is due to specific measurement errors of each support, whereas the other part represents the natural microstructure in which the regularization rules should be applied.

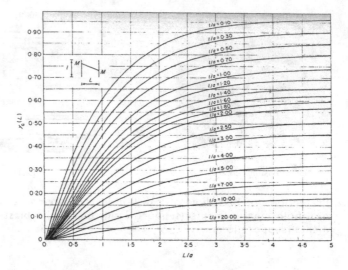

Figure 5.3 Example of nomogram to calculate regularized variograms, here an exponential model whose calculation direction is perpendicular to the direction of regularization. '*L*' is the calculation distance (to be standardized by
the parameter '*a*' of the model), and '*l*' is the size of the support. (Credit:
Jacques Laurent, draftsman of the Center of Geostatistics at Paris School
of Mines, end of the 1960s.)

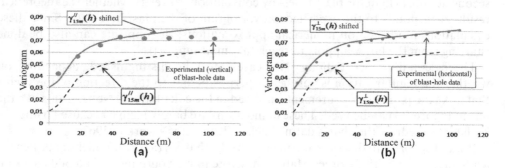

Figure 5.4 Deconvolution and convolution. Deconvolution of the variogram of the
drill-hole samples (3 m support), and then convolution to the support of
the blast holes (15 m). The result is presented in dotted lines; the continuous curve is its vertical shift of 0.02; the points are the experimental
variograms of the blast holes. (a) The calculation in the direction of regularization (vertical). (b) Calculation direction perpendicular to it.

To separate these two parts and to detect the amount that both types of measurement
share, a bivariable analysis is necessary. In this case of pure 'heterotopy' (no measurement of both types is present at exactly the same location), it would be necessary to
use centered covariances, the existence of which assumes second-order stationarity
that is not possible here due to the presence of a linear component (without a sill) in

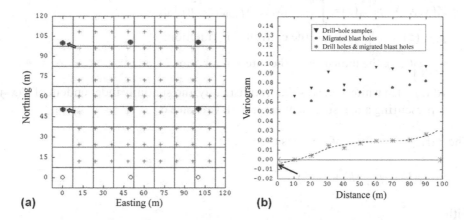

(a)   Easting (m)   **(b)**   Distance (m)

Figure 5.5 Data migration. (a) At each drill-hole sample (point), a blast-hole (cross) measurement is migrated when the distance does not exceed 10 m. (b) Triangles: experimental variogram of drill-hole samples; points: variogram of migrated blast holes; asterisks: cross-variogram between the two types of measurement, with a slight negative nugget effect. The near absence of a cross-nugget effect shows that the copper grades do not have a microstructure.

the variograms. Since the use of the covariance is not permissible, it seems better to calculate cross-variograms, which requires using the migration trick: move the measurement of the nearest blast hole to each drill-hole sample, as long as the distance does not exceed a threshold, as small as possible (10 m here), the one that allows obtaining sufficient measurement pairs (Figure 5.5a). This practice is debatable: if it were a true 'natural' nugget effect, by moving a measurement only 1 mm, any spatial correlation with its counterpart would be destroyed. But the true nugget effects are scarce. Will they exist in this type of deposit? The geologists' answer is negative: they sometimes observe microstructures with a range of a few meters. Since the migrations involve distances ranging from a few decimeters to a maximum of 10 m, with an average of about 5 m, a slight alteration of the correlations is expected, but not what is observed in Figure 5.5b: a drop to zero (or even to a slightly negative value) of the cross-correlation.

The model that emerges, therefore, is to give each type of measurement its own error:

$$Z_{\text{blast}}(x, y, z) = Z(x, y, z) * p_{15\text{m}}(z) + R(x, y, z) \tag{5.1}$$

with

- $Z_{\text{blast}}(x, y, z)$, the grade measured in the blast hole centered at the location with coordinates (x, y, z);
- $Z(x, y, z)$, the true point-support grade, without measurement error, at the location with coordinates (x, y, z);
- *, the convolution product;

- $Z(x,y,z) * p_{15m}(z) = \int_{-\infty}^{+\infty} Z(x,y,u)\, p_{15m}(z-u)\mathrm{d}u;$

- $p_{15m}(z) = \frac{1}{15} 1_{\left[0,\frac{15}{2}\right]}(|z|),$ the convolution function;

- $1_{\left[0,\frac{15}{2}\right]}(|z|),$ the indicator of the interval [−7.5 m, 7.5 m];

- $R(x, y, z)$, a white noise residual, statistically and spatially independent of $Z(x, y, z)$, representing a measurement error with variance σ_R^2.

The variogram of $Z_{blast}(x, y, z)$ is then:

$$\gamma_{blast}(h) = \gamma_{15m}(h) + \gamma_R(h) \tag{5.2}$$

with

- $\gamma_R(h)$, the variogram of the measurement error of the blast holes, assimilated to a nugget effect with variance σ_R^2;
- $\gamma_{15m}(h) = (\gamma * P_{15m})(h) - (\gamma * P_{15m})(0)$, the variogram regularized to a vertical support of 15 m;
- γ, the underlying point-support variogram;
- $P_{15m}(h)$ (with an uppercase P), the variogram regularization function, which is written as the self-convolution of p_{15m} (lowercase):

$$P_{15m}(h) = \left(p_{15m} * \breve{p}_{15m}\right)(h) = \frac{1}{15^2}\left(15 - |h_z|\right) 1_{[0,15]}(|h_z|) \text{ with } h = (h_x, h_y, h_z).$$

Similar equations are obtained for the drill-hole samples by replacing 15 m with 3 m and by introducing another residual $R'(x, y, z)$.

5.1.3 Predictions

Equations (5.1) and (5.2) allow performing interesting operations, such as filtering the measurement error by factorial kriging analysis (Matheron, 1982) or eliminating from the blast holes the effect of regularization by 'deconvolution' in order to predict a point-support value, an opportunity to update some little known works on how to increase the resolution of scanner images (Séguret, 1988a, b; Le Loc'h, 1990), works taken up and published by Jeulin and Renard (1992) and then by other authors such as Desnoyers and Dogny (2014).

The linear systems involved are, in general form:

$$\begin{pmatrix} \gamma_{15m} + \gamma_R & 1 \\ 1 & 0 \end{pmatrix} \begin{pmatrix} \lambda \\ \mu \end{pmatrix} = \begin{pmatrix} \gamma_{15m} + \sigma_R^2 \\ 1 \end{pmatrix} \tag{5.3}$$

$$\begin{pmatrix} \gamma_{15m} + \gamma_R & 1 \\ 1 & 0 \end{pmatrix} \begin{pmatrix} \lambda \\ \mu \end{pmatrix} = \begin{pmatrix} \gamma * p_{15m} - (\gamma * p_{15m})(0) + \sigma_R^2 \\ 1 \end{pmatrix} \tag{5.4}$$

In Eq. (5.3), which represents the filtering of the measurement error, the variogram of the residual γ_R disappears from the second member of a traditional kriging system and is replaced by the nugget effect variance σ_R^2, whereas in Eq. (5.4), which represents the deconvolution and elimination of the measurement error, the regularized variogram $\gamma_{15m}(h) = (\gamma * P_{15m})(h) - (\gamma * P_{15m})(0)$ of Eq. (5.2) that uses the self-convolution P_{15m} (uppercase) is replaced in the second member by $(\gamma * p_{15m})(h) - (\gamma * p_{15m})(0)$ (p_{15m}, lowercase and a single regularization).

The methods are tested on a simulation, considered as the reality, where all the quantities are known exactly, whether they are on a point support or regularized on a larger support. For simplicity, only the blast-hole data are affected by a measurement error, not the drill-hole samples. Figure 5.6c shows the result of a classical kriging that aims at predicting, at each measurement point, an underlying point-support value by using the data regularized at 15 m: the correlation with the reality is poor (0.55). When the measurement error is filtered by kriging (without deconvolution) using the system (5.3), the improvement is spectacular, and the correlation increases to 0.78 (Figure 5.6b). When the deconvolution is added by solving system (5.4), the performance improves again (Figure 5.6a, leading to a correlation of 0.85). Why?

Since the systems of Eqs. (5.3) and (5.4) correspond to a filtering, the set of points used locally in the prediction (the famous 'kriging neighborhood'; see Section A.3 in Appendix) can incorporate the point targeted for prediction; the latter does not receive a weight equal to 1, as it would be the case in the classical system of Figure 5.6c, for which the target point cannot be included. But this target point, due to its privileged position and in this context of filtering, receives a weight of around 0.65. Although noisy due to the measurement error, its value is closer to the reality than any average based on the surrounding points because it is in the exact location targeted for prediction. The real interest of these systems is to allow the target point to be incorporated into the system, which could give ideas for many applications.

Let us now address the true objective of the study, the use of data sets known on two different supports to predict the average value of a 3,000 m^3 block. The developed system is a cokriging with unknown but related means (Emery, 2012a), made possible by the equality between the average grades of drill holes and blast holes:

$$
\begin{pmatrix}
\gamma_{3m} & \gamma_{3m,15m} & 1 \\
\gamma_{3m,15m} & \gamma_{15m} + \gamma_R & 1 \\
1 & 1 & 0
\end{pmatrix}
\begin{pmatrix}
\lambda \\
\lambda' \\
\mu
\end{pmatrix}
=
\begin{pmatrix}
\gamma * p_{3m} * p_V - (\gamma * p_{3m} * p_V)(0) \\
\gamma * p_{15m} * p_V - (\gamma * p_{15m} * p_V)(0) + \sigma_R^2 \\
1
\end{pmatrix}.
$$

$$(5.5)$$

where p_V is the convolution function of block V. The specificity of the related means is extremely important: it gives an identical influence to both types of measurements. A classical ordinary cokriging system (with unknown and unrelated means) would force one of the two types to be chosen as auxiliary, which would considerably reduce its influence (Wackernagel, 2003). In the present case, the blast-hole and drill-hole measurements are of comparable quality; there is no reason to privilege one type of measurement more than the other, with greater reason because the links with their respective supports have been accounted for and formalized.

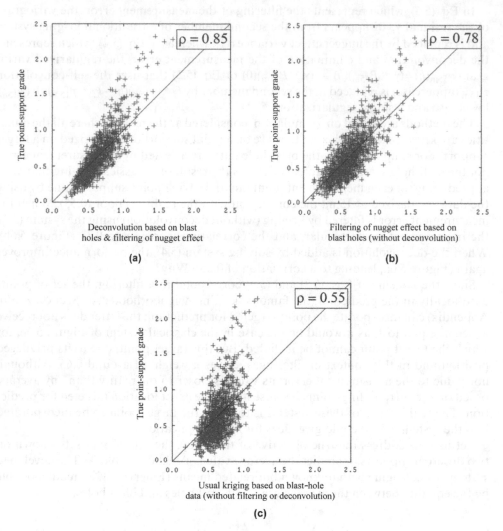

Figure 5.6 Deconvolution by kriging. The joint analysis of the grades from drill holes and blast holes shows coherence, in the sense that, except for the measurement errors, the algebraic link between these quantities responds to the regularization laws. With this acquired fact, some unusual calculations can be made, such as filtering the measurement error or predicting the grade on an infinitely small support ('point') from the known grades on a much larger support. To evaluate the performance of these calculations, a realistic geostatistical simulation on a fine mesh is used, which allows all the calculations to be made accurately and, therefore, to compare the predictors with the 'reality'. The cross-diagrams between the true point-support values (vertical axes) and their predictions (horizontal axes) are presented. All the systems use the known grades in blast holes. (a) Kriging with deconvolution and filtering of the measurement error; (b) kriging only with filtering; (c) classical kriging.

Figure 5.7a shows the comparison between the true block-support averages and their prediction by means of a kriging using the few drill-hole samples around the target. The result is poor, compared with the same kriging that uses the blast holes, four times more numerous (Figure 5.7b). The increase in the correlation with reality shows once again the importance of the information, which is so abundant here that the incorporation of the drill-hole samples to the blast-hole data barely changes the results (Figure 5.7c, cokriging).

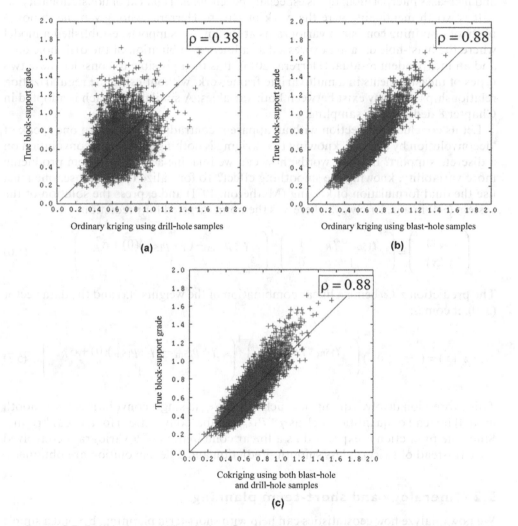

(a)

(b)

(c)

Figure 5.7 Drill holes and blast holes. Cross-diagrams between the true block-support grades (vertical axes) and their predictions (horizontal axes): (a) kriging using the drill-hole data; (b) kriging using the blast-hole data; (c) cokriging using blast-hole and drill-hole data (system (5.5)).

5.1.4 Discussion and perspective

Should one be disappointed by the lack of effectiveness of cokriging? Not really, because it has been tested in an unfavorable condition, where the drill-hole information is immersed among a dense group of blast holes. In practice, cokriging could be useful in an extrapolation context, in an area devoid of blast holes and where a few drill-hole measurements would exist, exactly the situation that occurs in short-term planning, when predictions are made below the depth of blast-hole sampling. Incorporating such drill-hole data, although scarce, can be extremely useful as it reduces extrapolation (known to be unrobust) and increases interpolation; this is especially beneficial in a context of nonstationarity.

It is worth mentioning here the work of Álvaro Herrera, who, given the impossibility of obtaining consistent variograms at the known supports, established a model where the blast-hole data are expressed as a linear combination of the drill-hole data and an independent residual (Herrera, 2013). It is then possible to consider these two types of measurements in a multivariate framework, without using the regularization relationships that may exist between both variables. A similar approach is adopted in Chapter 8 dedicated to sampling.

Let us conclude this section with an apparent contradiction: how can one speak of 'deconvolution by kriging', knowing that kriging is nothing else than a convolution on a discrete support? In other words, how can we imagine a kriging system producing more variability, knowing the smoothing effect? To formalize the response, one must use the dual formulation of kriging (Matheron, 1971) and express the solution of the system (5.4) in a generic way, where x is the point targeted for prediction:

$$\begin{pmatrix} \lambda(x) \\ \mu(x) \end{pmatrix} = \begin{pmatrix} \gamma_{15m} + \gamma_R & 1 \\ 1 & 0 \end{pmatrix}^{-1} \begin{pmatrix} \gamma * p_{15m} - (\gamma * p_{15m})(0) + \sigma_R^2 \\ 1 \end{pmatrix}. \tag{5.6}$$

The prediction $z^*(x)$ being a linear combination of the weights $\gamma(x)$ and the data vector $(z, 0)$, it comes:

$$z^*(x) = \begin{pmatrix} z & 0 \end{pmatrix} \begin{pmatrix} \gamma_{15m} + \gamma_R & 1 \\ 1 & 0 \end{pmatrix}^{-1} \begin{pmatrix} \gamma * p_{15m} - (\gamma * p_{15m})(0) + \sigma_R^2 \\ 1 \end{pmatrix}. \tag{5.7}$$

This expression involves quantities such as $\gamma * p_{15m}$ (a single convolution, less smooth model) instead of quantities such as $\gamma * P_{15m}$ (double convolution) for a usual kriging. Since the prediction is expressed as a linear combination of a variogram convolved once (instead of twice), a less smooth prediction and a deconvolution are obtained.

5.2 Mineralogy and short-term planning

We now analyze how geostatistics can help with short-term planning, beyond a simple mapping of block-support grades taking into account the differences in support of the exploration and production data. In the context of heap leaching for oxidized copper ore (see Chapter 7), mineralogy plays a key role, since small variations in the mineral abundances can have a great influence on the copper recovery and acid consumption, which directly affects the treatment costs. Likewise, in the case of the concentration of sulfide copper ore by flotation, the mineralogy determines the consumption of re-agents and the concentrate quality, in particular its copper grade and the grades of by-products and contaminants. In any case, the production geologists are interested in a predictive model of the mineralogy and in determining the risk of finding some undesirable minerals in excess of some threshold.

The following study is again from the Radomiro Tomic mine, focusing on the area of surface oxides where, for each blast hole, the production geologists measured the abundance of several minerals. Three of them are considered here: copper clay, ataca-mite and copper wad (an oxide or hydroxide of manganese and copper) (Cuadra and Rojas, 2001). The measurement corresponds to a semiquantitative visual assessment (often a multiple of 5% or 10%) of the abundance of each mineral in relation to all oxidized copper minerals. Ten classes of measurements are defined for atacamite and copper clay and three classes for copper wad (Table 5.1). Their experimental distribu-tions are represented by the histograms of Figure 5.8. Also note that the abundances are not independent: the abundance of atacamite has a correlation of 0.40 with that of copper wad and −0.44 with that of copper clay.

The objective of the study is not only to map the minerals abundances, which can be achieved by cokriging, but also to determine, for each mineral, the probability that the extracted material belongs to a given class. Such a probability is useful for defining the ore mixtures that will be sent to the heap leaching process to improve recovery and to minimize acid consumption costs.

The three discrete regionalized variables under study are subject to order relation-ships between classes: provided that there is some spatial continuity, two adjacent lo-cations should belong to the same class or to two contiguous classes. Therefore, the study is oriented toward a truncated Gaussian model, which has already been applied in Chapter 1 for facies simulation, since this model allows obtaining variables distrib-uted in discrete and ordered classes.

Three stationary Gaussian random fields are considered, and for each of them, a set of truncation thresholds is defined in such a way that the truncated random fields present the distributions of Figure 5.8. Their direct and cross-variograms are then fitted with a linear model of coregionalization that includes a nugget effect and spherical structures with geometric or zonal anisotropy, so as to reproduce the di-rect and cross-variograms of the truncated variables (Figure 5.9). Note that the model

Table 5.1 Coding the abundances of three oxidized copper minerals (copper clay, atacamite and copper wad) in discrete classes

Class	Copper clay (%)	Atacamite (%)	Copper wad (%)
0	0–10	0–10	0–4
1	11–20	11–20	5–9
2	21–30	21–30	>10
3	31–40	31–40	-
4	41–50	41–50	-
5	51–60	51–60	-
6	61–70	61–70	-
7	71–80	71–80	-
8	81–90	81–90	-
9	91–100	91–100	-

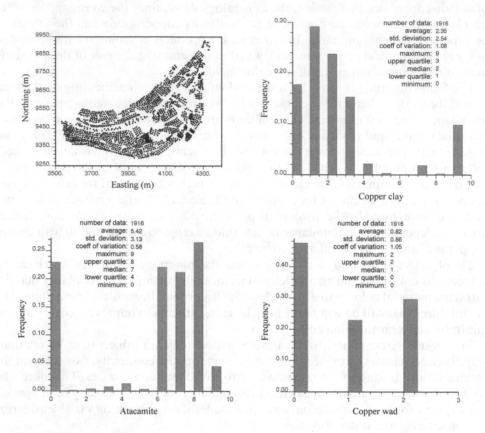

Figure 5.8 Map and histograms of mineralogy data for the blast holes located on the bench of elevation 2,825 m.

reproduces the positive correlation between the abundances of atacamite and copper wad and the negative correlation between the abundances of atacamite and copper clay; the correlation between the abundances of copper clay and copper wad is low.

Conditioned to the blast-hole data, the truncated Gaussian simulations allow mapping the spatial variability of mineral abundances (Figure 5.10), as well as the probabilities that these abundances are greater than given thresholds (Figure 5.11). In particular, the geologists are interested in the probability of excess of two minerals:

- the copper clay, which delays the percolation of the leaching solution and creates preferential channels and/or nonleached areas in the heap;
- the copper wad, which negatively affects the kinetics and the extraction rate of the leaching process.

The calculated probability maps (Figure 5.11) allow identifying high-risk areas, such as the northeast zone for copper clay or the south–southeast zone for copper wad. The ore extracted from these areas must be mixed with the material extracted from low-risk areas, so a stockpile management function is necessary.

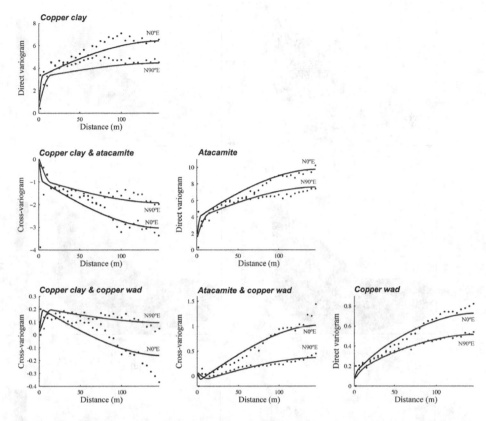

Figure 5.9 Direct and cross-variograms of the mineral abundances encoded in discrete classes, calculated along the main directions of anisotropy (N0°E and N90°E). Points: experimental variograms. Continuous lines: modeled variograms (Emery and Cornejo, 2010).

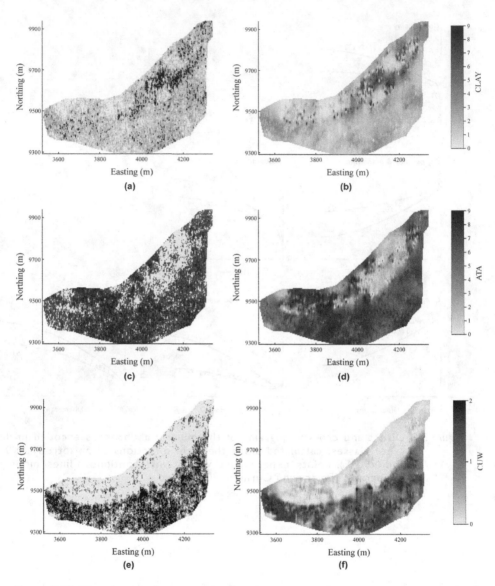

Figure 5.10 Abundance of copper clay (a, b), atacamite (c, d) and copper wad (e, f). *Left*: simulation n°1. *Right*: average of 100 simulations. The abundances are cosimulated on a regular grid with mesh 4 × 4 m² for the bench of elevation 2,825 m.

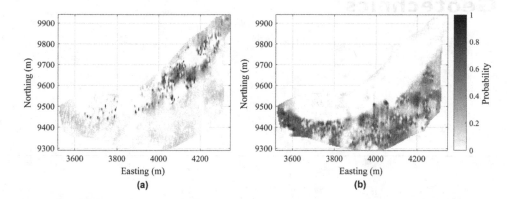

Figure 5.11 Maps of the probability of exceeding 50% of copper clay (a) or 5% of copper wad (b), calculated from 100 simulations. High probability areas present a risk of excess of copper clay or copper wad, which negatively affects the heap leaching process.

Chapter 6

Geotechnics

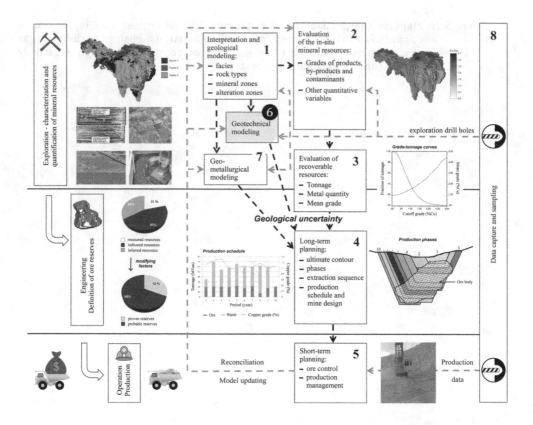

Mining geotechnics, a discipline at the intersection of geology, geophysics and geomechanics, is a formidable opportunity for geostatistics: almost everything remains to be done. To determine the quality of a rock mass, the usual practice is to divide the deposit into domains called 'geotechnical units' and to characterize them by an average value. This practice is debatable as, with some exceptions such as the one that is studied in the first section of this chapter, geotechnical variables are often nonadditive, nonlinear by change of support, and directional, that is, the measured value depends not only on the position of the measurement but also on its direction. Therefore, one should interpolate in space not only numerical quantities but also vectors, or even tensors, which is not formally done at present. But these difficulties can be overcome. The physics that governs rock mechanics is so powerful that it makes the laws of statistics emerge, upon which a geostatistical modeling can be based, and this leads to surprising methods. We hope to demonstrate these in this chapter.

The following sections present geostatistical studies of

- the discontinuity intensity (P_{32}), an additive variable that quantifies the degree of fracturing of the rock mass based on the area covered by discontinuities per unit volume of rock;
- the linear fracture frequency (FF), which relies on a count of discontinuities;
- the rock quality designation (RQD);
- the link between these last two directional and nonadditive variables.

Along the way, a new concept is introduced, the 'directional concentration', a measure of the tendency of the discontinuities to align themselves at a small scale. This variable leads to a new attribute that quantifies the solidity of the rock: the *crushing*.

The last part of this chapter focuses on the concept of directionality (Section 6.5) and presents what is likely to be the ultimate solution to this problem (Section 6.6): a regionalization of directional variables in a 5D spatioangular space.

6.1 Discontinuity intensity

To date (2020), most of published geostatistical works applied to the study of rock mass fracturing involve the modeling of 'discontinuities' – a generic term to designate rock ruptures with or without displacement, such as faults, fractures, joints, veins, fissures – and the simulation of discontinuity networks (Chilès, 1989a, b; Chilès and de Marsily, 1993). In this context, an important task is the three-dimensional characterization of the discontinuity properties, including the definition of families of discontinuities and, for each family, the number, orientation, spacing, location, shape and size of the discontinuities. This characterization is based on three main sources of information:

- geotechnical drill holes, with or without oriented cores, which provide one-dimensional samples and petrophysical records;
- two-dimensional observation windows on outcrops or underground galleries, which allow, in particular, evaluating the size of the discontinuities via the study of the lengths of their traces (Baecher and Lanney, 1978; Warburton, 1980; Priest and Hudson, 1981);
- remote sensing that provides three-dimensional information (Sturzenegger et al., 2011; Riquelme et al., 2015; Tuckey and Stead, 2016).

6.1.1 Definition of the discontinuity intensity

The discontinuity intensity can be measured by several variables, represented under the generic notation P_{xy}, where x is the dimension of the sample (1, 2 or 3, depending on whether it is a line, a surface or a volume) and y is the dimension of the measurement

(0, 1, 2 or 3, depending on whether it is a count, a line, a surface or a volume). In particular, P_{10} coincides with the linear FF (fracture frequency) that is studied in Section 6.2. For now, our attention is focused on P_{32}, the average area of discontinuities per unit volume of rock mass (Dershowitz and Herda, 1992). This is one of the preferred variables of the geologists and geotechnicians of the El Teniente underground mine, since it is additive in the sense that, if several disjoint blocks of equal volume are considered, the discontinuity intensity of the union of these blocks is equal to the arithmetic mean of the discontinuity intensities measured on the blocks.

6.1.2 Problem and data

The rock mass of the El Teniente mine includes several rock types: the Braden pipe (a waste breccia) that was modeled in Chapter 1, anhydrite breccias, mafic intrusive rocks (gabbro, diabase and El Teniente mafic complex or CMET, for its acronym in Spanish) and felsic intrusive rocks (dacite and dioritic to tonalitic porphyry) (Skewes et al., 2006).

Of the two main types of discontinuities observed in the ore extracted from the mine – large-scale faults and swarms of small veins (stockwork) filled with a weak mineral assemblage, called 'weak veins' – the geologists and geotechnicians of El Teniente have found that the latter are the most important factor in explaining the fragmentation of the rock and the geomechanical instability of the mine (rock bursts) (Figure 6.1) (Brzovic and Villaescusa, 2007; Brzovic, 2009; González and Brzovic, 2015).

CMET rock geotechnical data collected by a team of geologists from the El Teniente mine between 2010 and 2016 are available for this study. The data contain the position of the weak veins that intersect each core and their orientation with respect to the axis of the drill hole. Two sectors are considered and modeled separately, the 'East-CMET' and the 'West-CMET', separated by the Braden pipe and the dacite (Figure 6.2).

Figure 6.1 Stockwork or swarm of veins and veinlets (straight and curvilinear structures in white and light gray) in the primary ore, observed on a mine drive of the El Teniente mine. These veins have a lower resistance than the intact rock and influence the stability and fragmentation of the rock mass. (Credit: Andrés Brzovic and Paulina Schachter.)

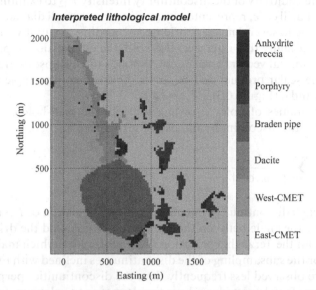

Figure 6.2 Lithological model of El Teniente for elevation 2,150 m. The Teniente mafic complex (CMET) covers most of the region and is separated into two sectors (East-CMET and West-CMET) by the Braden pipe (circular body studied in Chapter 1) and the dacite (body elongated along the north–northwest direction).

6.1.3 Methodology

Its additive character makes the discontinuity intensity P_{32} a 'nice' variable, since it can be modeled and interpolated in the same way as a copper grade. However, the difficulty here is the absence of direct measurements of P_{32}, since there is only a record of the weak veins that intersect a geotechnical drill hole and their orientations relative to the drill-hole axis.

To address this difficulty, the geologists of the mine usually predict P_{32} from FF, first by simulating numerous networks of weak veins, then selecting the simulations that best match the FF values observed along the drill holes, and finally calculating a factor (C_{31}) to convert FF into P_{32} in the selected simulations (Dershowitz and Herda, 1992). All this is very laborious and, above all, subject to possible biases. In fact, as will be seen in the following, the FF and the conversion factor C_{31} are nonadditive quantities, subject to sampling biases and prediction biases.

Another approach is to estimate the number of weak veins per unit volume of rock and their diameter distribution, assuming that such weak veins can be represented by discs whose positions, orientations and diameters are random (Zhang and Einstein, 2000; see also Section A.5.2.3 in Appendix). Both sources of information allow estimating the distribution of the discontinuity intensity P_{32}, which corresponds to the average area covered by veins per unit volume of rock. Unfortunately, both the estimation of the number of veins and the estimation of their diameter distribution are generally not robust, as pointed out by Chilès et al. (2008). These authors advocate a direct approach that is explained and used in the following.

Let us use the additivity of the discontinuity intensity P_{32} to examine this variable at the support of a drill core, represented by a cylinder v of small diameter δ (a few inches) and length L (one to several meters), oriented according to a unit vector α. When considering a discontinuity that intersects this core, the intersection, represented by a flat surface of unit normal vector (pole) ω, corresponds to an ellipse of area $\pi\delta^2/(4|<\omega, \alpha>|)$, where $<,>$ is the scalar product and $<\omega, \alpha>$ is the cosine of the angle between the two unit vectors ω and α (Figure 6.3).

If N discontinuities of poles $\omega_1 \ldots \omega_N$ intersect the drill-hole core (a cylinder of volume $\pi\delta^2 L/4$), the discontinuity intensity for this core is:

$$P_{32}(v) = \frac{1}{L} \sum_{i=1}^{N} \frac{1}{|<\omega_i, \alpha>|}. \tag{6.1}$$

Accordingly, each discontinuity contributes to the calculation of P_{32} by the reciprocal of the cosine of the angle between the discontinuity pole and the drill-hole axis. It is nothing less than the Terzaghi correction (Terzaghi, 1965), which makes it possible to compensate for the subsampling of the discontinuities inclined with respect to the drill hole, which are observed less frequently than the discontinuities perpendicular to the drill hole. Formula (6.1) provides a very simple means to calculate, from the observations recorded in the database, the discontinuity intensity P_{32} for a cylindrical support of small diameter and length L (here, $L=1$ m). In practice, a numerical correction is necessary if the relative angle is close to 90°, since it would then be divided by 0 in formula (6.1). Too large angles must be decreased to a maximum value, for example, 75° (Chilès et al., 2008).

The calculated P_{32} data (Figure 6.4) can be treated exactly as if they were copper grade measurements, opening the way for kriging or for conditional simulation techniques that are now applied.

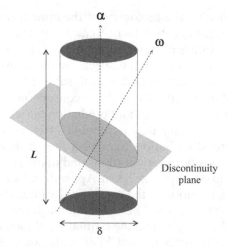

Figure 6.3 Intersection between a flat discontinuity and a drill-hole core.

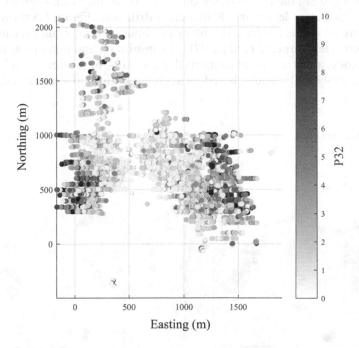

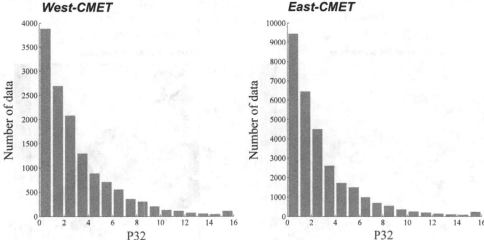

Figure 6.4 Map (projection on a horizontal plane) of the calculated P_{32} data for drill-hole cores of I m in length, and histograms distinguishing the West-CMET and East-CMET sectors.

6.1.4 Results

The modeling of the data in each sector is not detailed, as it follows the classical steps for simulation under the multi-Gaussian framework, already explained in Chapters 2 and 4. A total of 500 simulations are constructed, producing P_{32} values on a regular grid of $4 \times 4 \times 10$ m^3, then regularized on a $20 \times 20 \times 20$ m^3 block support and restricted to the blocks distant less than 100 m from a drill hole. Figure 6.5 shows one of these simulations, as well as the block-by-block minimum, average and maximum of the 500 simulations (Hekmatnejad et al., 2017). The minimum and maximum values provide, for each block, an interval that contains the true value of the weak vein intensity with a confidence of $499/501 = 99.6\%$, based on arguments given in Section A.4.5 (Appendix).

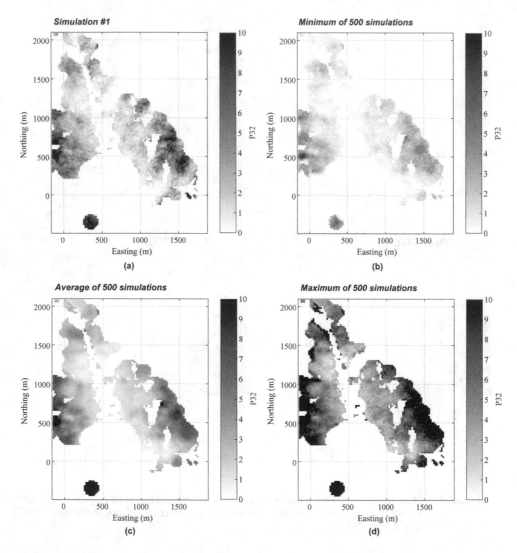

Figure 6.5 Horizontal cross-sections of P_{32} at elevation 2,150 m: (a) one of the 500 simulations of P_{32} on a block support of $20 \times 20 \times 20$ m^3, (b) block-by-block minimum, (c) average and (d) maximum of 500 simulations. Each simulation has more variability than the average of the simulations (smoothed) but less than the original data, due to the change of support between the drill-hole cores and the simulated blocks.

An interval associated with any level of confidence can be derived by considering the distribution of the 500 simulated values.

Some sectors of low vein intensity are adjacent to the dacite, whereas the high vein intensity sectors are located in the eastern and western sides of the study area, as well as to the southwest of the Braden pipe, but the latter sector contains very few drill holes. The maps in Figure 6.5 are useful for identifying the geotechnically most dangerous sectors.

6.1.5 Discussion

The discontinuity intensity is an exceptional case in mining geotechnics, as this variable is additive and can be interpolated by kriging or conditional simulation techniques without any difficulty. The absence of direct measurements is only apparent, since the discontinuity intensity can be calculated for any drill-hole core sample from the count of the discontinuities that intersect this sample and their orientations with respect to the drill-hole axis (Eq. 6.1), information that is available in geotechnical databases.

6.2 Fracture frequency

At the end of 2011, we were asked to study the variations in space of a nonadditive attribute, the Rock Mass Rating (RMR, Barton et al., 1974), a concatenation of several variables in a table used by the engineer to evaluate the rock quality and to design the ground support. To understand RMR, one needs to study its constitutive variables, among which the FF and the RQD, two variables related to fracturing, which will be quantitatively studied without the restriction of a genetic model aimed at reproducing the discontinuities in space.

The literature presents some geostatistical studies of these variables (La Pointe and Hudson, 1985; Barton and La Pointe, 1995), which always focus on the count of the fractures on supports with a constant size. However, in Chuquicamata and Radomiro Tomic, we faced a serious problem that persists today: crushing. Some pieces of the drill-hole cores, whose lengths change from one core to another, are fragmented and cannot be used to count the fractures. This makes the measurement of FF nonadditive because the fractures are counted in the intact pieces of variable lengths, although all the original core samples have the same length (1.5 m in this case).

The starting point of the study was to observe the practical formula used by our Chilean collaborators and to interpret its hidden meaning. The following step was to consider the crushing phenomenon not as an issue that should be eliminated quickly by a magic formula, but as an advantage. Seen with the formalism of regionalized variables, crushing appeared as interesting as the number of fractures, since it is an intrinsic characteristic of the rock, sometimes related to the number of fractures and sometimes not. The statistical link depends on a hidden variable that had to be constructed: the 'directional concentration', a measure of the tendency of the fractures to be, at the scale of the drill hole cores, parallel to each other or, conversely, to cross, and to act as a natural crusher that fragments the rock.

Instead of a variable FF equal to the number of fractures divided by the intact portion of the drill-hole core, that is, a nonadditive ratio, which is furthermore too synthetic and difficult to interpret, the modeling then focused on a vector of three regionalized variables – the number of fractures, the fragmented core length and the directional concentration. Since the statistical correlation of the first two variables is a function of the third one, this correlation also becomes a regionalized variable for modeling, allowing mapping areas of space that are more or less fractured than it seems when only the FF is considered.

6.2.1 About the nonadditivity of fracture frequency

Figure 6.6 presents the elements of the problem: in each sample of constant length L, only the intact part L_{NC} allows the number of fractures to be counted.

To calculate the linear FF introduced by John Jaeger and Neville Cook in 1969, in a book that still serves as a reference today, one must divide the number of fractures N_{tot} by the analyzed (noncrushed) length L_{NC}:

$$FF_{true}(x) = \frac{N_{tot}(x)}{L_{NC}(x)}. \tag{6.2}$$

This formula poses a problem for spatial prediction, since the fragmented core length L_C changes from one core to another. It is not satisfactory to perform a kriging by putting together the measurements selected around the target point to calculate an optimized average, if these measurements are not comparable between them. In other words, the regionalized variable is not additive. How can one then predict the quotient defined in Eq. (6.2)? The answer will be to generalize a trick used in a phosphate deposit in northern Saudi Arabia a few years before this study: looking for an unbiased predictor that is not necessarily optimal (Chilès and Séguret, 2009). In the case of phosphate, the service variables were the accumulation, the thickness and the grade. There is a link between these variables, since the accumulation is a combination of the

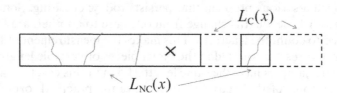

$L_{NC}(x)$: total noncrushed length of the sample located at x

$L_C(x)$: total crushed length of the sample located at x

$L_{NC}(x) + L_C(x) = L$ (constant) $\forall x$

$N_{tot}(x)$: total number of fractures along L_{NC}

Figure 6.6 Counting and crushing. Of the total length L, only L_{NC} allows counting the fractures, the complement being fragmented or disaggregated ('crushed'). How to map FF when L_{NC} changes from one core to another? A problem that is, for example, not present in El Teniente where little altered rocks are mined. But here in Chuquicamata and Radomiro Tomic, the problem is crucial.

grade and the thickness; by using this link under certain restrictions, the unbiasedness of the quotient prediction is immediately proven. But this direct relationship between numerator and denominator is not indispensable. We show it in the following, with greater reason because the proof contains a lesson about additivity that will be used in the next chapter dedicated to geometallurgy.

Consider two random fields Z_1 and Z_2 that are second-order stationary, whose quotient must be evaluated at a location x using a set of $2n$ measurements $\{z_1(x_i), z_2(x_i): i=1... n\}$ (isotopic sampling). Also suppose that these random fields are in intrinsic correlation, i.e., their direct and cross-variograms are proportional to each other, and that the ratio Z_1/Z_2 is independent of Z_2. Under these assumptions, the ratio between the kriging of Z_1 and the kriging of Z_2 is an unbiased predictor of the ratio Z_1/Z_2.

Indeed, because Z_1 and Z_2 are in intrinsic correlation and are known at the same data points, their krigings use the same weights $\{\lambda_i: i=1... n\}$. Hence, the quotient of these krigings is:

$$\frac{Z_1^*(x)}{Z_2^*(x)} = \frac{\sum_{i=1}^{n}\lambda_i Z_1(x_i)}{\sum_{j=1}^{n}\lambda_j Z_2(x_j)} = \frac{\sum_{i=1}^{n}\lambda_i Z_2(x_i)\dfrac{Z_1(x_i)}{Z_2(x_i)}}{\sum_{j=1}^{n}\lambda_j Z_2(x_j)} = \sum_{i=1}^{n}W_i \frac{Z_1(x_i)}{Z_2(x_i)} \qquad (6.3)$$

with

$$W_i = \frac{\lambda_i Z_2(x_i)}{\sum_{j=1}^{n}\lambda_j Z_2(x_j)} \text{ and } \sum_{i=1}^{n}W_i = 1. \qquad (6.4)$$

Knowing the values of Z_2 at the measurement locations has no influence on the distribution of the ratio Z_1/Z_2 due to the assumption of independence between Z_2 and Z_1/Z_2, while the random variable W_i conditional on the Z_2-values becomes a deterministic quantity denoted as w_i, whose expression is given by Eq. (6.4) by replacing the random field Z_2 (uppercase) with the deterministic field z_2 (lowercase). Therefore, one has:

$$E\left\{\frac{Z_1^*(x)}{Z_2^*(x)}\bigg| Z_2(x_i) = z_2(x_i): i=1...n\right\} = E\left\{\sum_{i=1}^{n}w_i \frac{Z_1(x_i)}{Z_2(x_i)}\right\}$$

$$= \sum_{i=1}^{n}w_i E\left\{\frac{Z_1(x_i)}{Z_2(x_i)}\right\}$$

$$= E\left\{\frac{Z_1(x)}{Z_2(x)}\right\}. \qquad (6.5)$$

Randomizing z_2 to obtain the nonconditional expectation, it comes:

$$E\left\{\frac{Z_1^*(x)}{Z_2^*(x)}\right\} = E\left\{E\left\{\frac{Z_1^*(x)}{Z_2^*(x)}\bigg|Z_2(x_i):i=1...n\right\}\right\} = E\left\{\frac{Z_1(x)}{Z_2(x)}\right\}. \qquad (6.6)$$

Accordingly, the expectation of the quotient between the kriging of Z_1 and that of Z_2 coincides with the expectation of the quotient Z_1/Z_2: the quotient of the kriging predictors is unbiased.

Do the aforementioned assumptions apply to the numerator and denominator of Eq. (6.2)? On the one hand, the independence between L_{NC} and N_{tot}/L_{NC} is corroborated by the fact that these variables are not correlated, since the average of their product is practically equal to the product of their averages. On the other hand, the intrinsic correlation assumption is also verified: the direct and cross-variograms of N_{tot} and L_C are proportional to each other (Figure 6.7), which is also true for the variograms of N_{tot} and L_{NC}. Although this intrinsic correlation means that the cokriging between the two variables is not helpful, it allows obtaining a method for predicting FF that, although not optimal, is unbiased, which would no longer be ensured in a context of nonintrinsic correlation.

6.2.2 The user's formula

Knowing very well that FF_{true} is not additive and ignoring the previous properties, geotechnicians used their common sense and made calculations involving the number of fractures and the fragmented length. They mapped the variable $FF_{corrected}$ defined as follows:

$$FF_{corrected}(x) = \frac{N_{tot}(x) + a L_C(x)}{L}. \qquad (6.7)$$

Having understood the importance of crushing, geotechnicians included the crushed length L_C in the definition of FF. In Eq. (6.7), $a=40$, why? And what is the meaning of this formula?

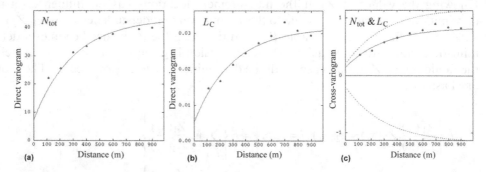

Figure 6.7 Experimental and modeled variograms. Points: experimental variograms; continuous lines: models. (a) Number of fractures; (b) fragmented core length (variogram identical to variogram of the intact length); (c) cross-variogram. The correlation is intrinsic, since the direct and cross-variograms are proportional to each other.

Rewriting Eq. (6.7) to make Eq. (6.2) appear, one gets:

$$
\begin{aligned}
FF_{corrected}(x) &= \frac{L_{NC}(x)FF_{true}(x) + L_C(x)\,a}{L_{NC}(x) + L_C(x)} \\
&= \frac{L_{NC}(x)FF_{true}(x) + L_C(x)\,FF_{crushed}(x)}{L_{NC}(x) + L_C(x)}
\end{aligned}
\tag{6.8}
$$

Then, Eq. (6.8) turns out to be a typical additive formula that combines two frequencies in proportion to the lengths they represent. The coefficient 'a' appears as an arbitrary frequency $FF_{crushed}$ associated with the crushing phenomenon. Complementary studies (Séguret et al., 2014) show that, to be consistent with the measurements information, a should be greater than 85, with a large uncertainty about the value to be adopted, between 80 and 120. However, the maps in Figure 6.8 show that the results change a lot with this parameter and also give results very different from the map obtained by the ratio (6.3) of the kriging predictors. The calculation (6.8) is correct but inaccurate and finally not very informative.

6.2.3 Searching for the hidden variable

Let us tackle the problem at its source by going to the warehouse where the drill-hole cores are stored.

What do we observe (Figure 6.9)? When there are few fractures, the length of fragmented core is small and vice versa. Is it a general law? Consider the scatter plots in Figure 6.10 for the Chuquicamata and Radomiro Tomic mines. In both cases, the number of fractures tends to increase with the fragmented core length, although the analyzed complementary length decreases, which shows the strength of the phenomenon and the difficulty of the fracture counting when the intact core length only measures about 10 cm. However, these scatter plots are far from those expected to model

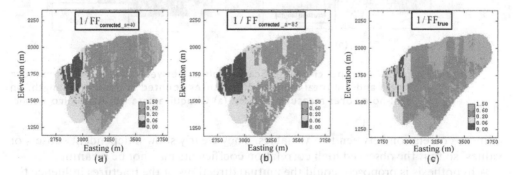

Figure 6.8 Fracture frequency (FF). Frequency inverses are represented in a vertical section XZ because the average length of the intact core pieces is more explicit than a frequency. The samples here measure 1.5 m. (a) Result for $FF_{corrected}$, formula (6.7) with $a = 40$; (b) the same with $a = 85$, the recommended value after the study; (c) the quotient of kriging predictors of formula (6.3).

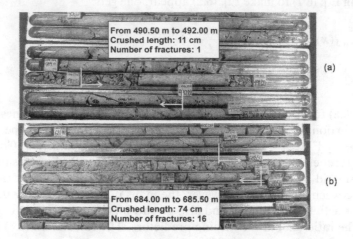

From 490.50 m to 492.00 m
Crushed length: 11 cm
Number of fractures: 1

(a)

(b)

From 684.00 m to 685.50 m
Crushed length: 74 cm
Number of fractures: 16

Figure 6.9 The crushing phenomenon. This study results from interrogations related to the observation of drill cores. (a) Few fractures, little crushing; (b) the opposite. (Credit: Cristian Guajardo.)

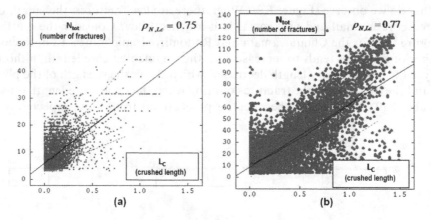

Figure 6.10 Fracturing and crushing. Scatter plots between the fragmented core length and the total number of fractures counted in the intact length, on drill-hole cores measuring 1.5 m. (a) Chuquicamata. (b) Radomiro Tomic.

a linear correlation between two variables. They clearly show independent subsets of values, so that the observed high correlation coefficients may not be meaningful.

A hypothesis is proposed: could the mutual directions of the fractures influence the disaggregation (crushing) of the core? If the fractures tend to cross, could they act as a natural crusher? To measure such a tendency, the direction of each fracture must be known. Fractures are classified according to their angle θ (the range of which can be discretized into n_θ classes) with respect to the core axis (Figure 6.11). The number of fractures can be written as follows:

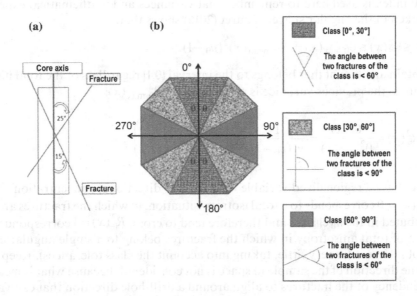

Figure 6.11 Classification of fractures in three classes. The core can be turned around
its axis to better examine a particular fracture, so the boundaries of
the angular classes are cones of revolution. (a) Example of two fractures. As the core sample is not oriented, the two fractures are placed
in the same class [0°, 30°] because only the smallest angle is considered.
(b) In terms of the potential to generate crushing, the classes are not
equivalent: classes 1 and 3 may contain fractures forming an angle of 60°
between them, whereas in the intermediate class, the fractures may be
perpendicular. But this is not really a limitation as, for this classification, it is enough that two of the three classes discriminate the extreme
situations – fractures approximately parallel or perpendicular to the core
axis – to obtain a useful result.

$$N_{\text{tot}}(x) = \sum_{\theta=1}^{n_\theta} N(\theta, x), \tag{6.9}$$

where $N(\theta, x)$ is the number of fractures belonging to class θ that intersect the sample
located at x, weighted by $1/\sin(\theta)$ in order to compensate for the subsampling of the
fractures that are inclined with respect to the drill hole (Terzaghi correction similar to
that made in Eq. 6.1 to calculate P_{32}). This sample located in x can be associated with
a random variable that acts as a variance, which can be calculated approximately by a
discrete sum based on the n_θ available classes:

$$\theta_\theta^2(x) = \text{var}_\theta[N(\theta, x)] = E_\theta[(N(\theta, x) - N_{\theta, \text{mean}}(x))^2]$$

$$\approx \frac{1}{n_\theta} \sum_{\theta=1}^{n_\theta} \left(N(\theta, x) - \frac{N_{\text{tot}}(x)}{n_\theta} \right)^2 \tag{6.10}$$

The θ index is used here to remember that variances and mathematical expectations are taken in the angular space. Séguret (2016) shows that:

$$0 \le \theta_\theta^2(x) \le \theta_{\theta,\max}^2(x) = N_{\theta,\text{mean}}(x)^2(n_\theta - 1). \tag{6.11}$$

To obtain an amount that belongs to the interval [0,1] regardless of the total number of fractures, the previous variance is normalized by $\theta_{\theta,\max}^2(x)$:

$$R_\theta^2(x) = \frac{\theta_\theta^2(x)}{\theta_{\theta,\max}^2(x)} = \frac{1}{n_\theta(n_\theta - 1)} \sum_{\theta=1}^{n_\theta} \left(\frac{N(\theta,x)}{N_{\theta,\text{mean}}(x)} - 1 \right)^2. \tag{6.12}$$

This defines a regionalized variable $R_\theta^2(x)$ called 'directional concentration'.

$R_\theta^2(x) = 0$ corresponds to a total isotropy situation, in which the fractures are equally distributed in the n_θ classes and therefore tend to cross. $R_\theta^2(x) = 1$ corresponds to a situation of total anisotropy in which the fractures belong to a single angular class and do not intersect (or very little, taking into account the class tolerances). Keep in mind that the direction of the sample in space is not considered, because what is measured is the tendency of the fractures to align around a drill-hole direction that can vary from one point to the other. The mapping of this directional concentration reveals large areas that correspond to particular rock types (Figure 6.12).

When the scatter plot between $L_C(x)$ and $N_{\text{tot}}(x)$ is now examined by classes of samples related to their directional concentrations, it is clear that when passing from

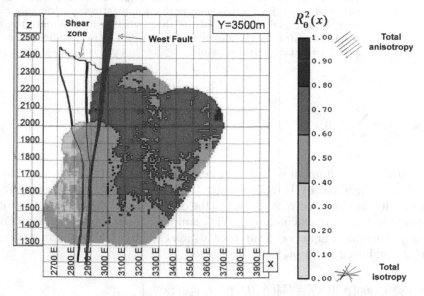

Figure 6.12 Chuquicamata. Vertical section of the directional concentration around the 'West Fault'. To the northeast of the fault, the fractures tend to be parallel to each other, unlike the southwest where they cross systematically, in an area where the rock is known to be of poor quality, extremely friable.

$R_\theta^2(x) = 0$ to $R_\theta^2(x) = 1$, the linear correlation coefficient goes from more than 0.8 to less of 0.2, and this applies to both Radomiro Tomic (Figure 6.13) and Chuquicamata. The nonelliptical nature of the plots in Figure 6.10 is explained by the fact that they gather samples for which, sometimes, there is no correlation between fracturing and crushing, which is explained by $R_\theta^2(x)$. When the fractures intersect, they produce disaggregation; when they tend not to cross, crushing is independent of fracturing and constitutes an intrinsic property of the rock that must be studied in the same way as fracturing.

6.2.4 Regional linear correlation model

In summary, for each 1.5 m sample located at point x, three coregionalized variables, associated with as many random fields, are available:

- the total number of fractures $N_{tot}(x)$;
- the fragmented core length $L_C(x)$;
- the directional concentration $R_\theta^2(x)$.

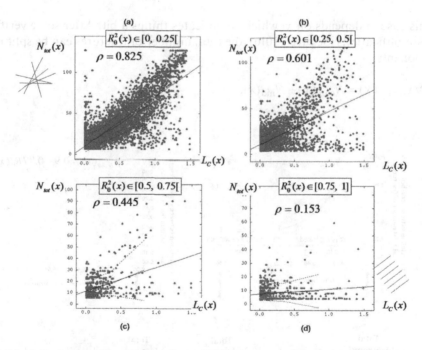

Figure 6.13 Radomiro Tomic data. Scatter plot between the fragmented core length $L_C(x)$ (horizontal axis) and the total number of fractures $N_{tot}(x)$ (vertical axis). Selected samples have a directional concentration (a) between 0 and 0.25 (b) between 0.25 and 0.5, (c) between 0.5 and 0.75, (d) between 0.75 and 1. When the fractures intersect often (cases (a) and (b)), they produce disaggregation; when they tend not to cross (cases (c) and (d)), crushing is independent of fracturing and constitutes an intrinsic property of the rock that must be studied in the same way as fracturing.

Since the linear correlation coefficient between the first two variables is a function of the third, it is possible to convert the information contained in $R_\theta^2(x)$ in a regionalized correlation coefficient by means of a nonlinear transformation:

$$\rho_{N,LC}(x) = f(R_\theta^2(x)). \tag{6.13}$$

The shape of function f depends on the studied deposit: exponential for Chuquicamata (Figure 6.14a) and linear for Radomiro Tomic (Figure 6.14b). In the first case, the curve starts arbitrarily from 1, which is the maximum correlation in the case of total isotropy, whereas in the second case, a maximum correlation equal to 0.9 is objectively found.

When two random fields Z_1 and Z_2 are related by a linear correlation coefficient ρ, one of the two (for example, Z_2) can be expressed linearly as a function of the other plus a residual ε statistically uncorrelated with Z_1, even spatially uncorrelated with Z_1 if the direct and cross-variograms of Z_2 and Z_1 are proportional (Rivoirard, 1994):

$$Z_2(x) = \frac{\sigma_2}{\sigma_2}\left(\rho Z_1(x) + \sqrt{1-\rho^2}\,\varepsilon(x)\right) + c. \tag{6.14}$$

In this case, ρ depends on x, which complicates things a bit. After some verifications and simplifications (Séguret, 2016), the total number of fractures can be split into two components:

$$\forall x, N_{\text{tot}}(x) = N_{\text{corr}}(x) + N_{\text{ind}}(x). \tag{6.15}$$

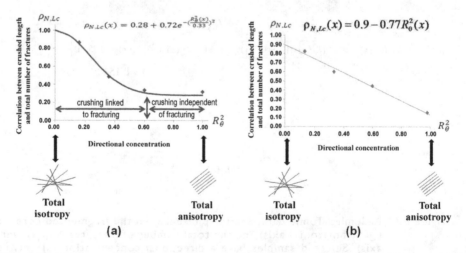

Figure 6.14 Linear correlation coefficient depending on the directional concentration. (a) Chuquicamata: for a concentration greater than 0.6, the correlation no longer changes, which defines a class of samples (and rock) where fracturing and crushing are independent. (b) Radomiro Tomic: fracturing and crushing are linked in the full range of directional concentration.

$N_{corr}(x)$ represents the number of fractures that is expected on average, knowing the fragmented core length $L_C(x)$ and the regional correlation $\rho_{N,Lc(x)}$. $N_{ind}(x)$ represents the number of fractures that must be added to $N_{corr}(x)$ to obtain the truly measured amount $N_{tot}(x)$. If $N_{ind}(x)$ is negative, then fewer fractures have been measured than those predicted when considering $L_C(x)$ and $\rho_{N,Lc}(x)$. If, on the other hand, $N_{ind}(x)$ is positive, a greater number of fractures than expected are counted.

Do such calculations make sense?

Figure 6.15a shows the number of independent fractures N_{ind} in a vertical section of Chuquicamata, near the West Fault. In the area known to be damaged to the west, fractures are much more numerous than they should be. The other two figures (Figure 6.15b and c) refer to the Radomiro Tomic deposit. Figure 6.15b shows a horizontal section of FF_{true} as defined in Eq. (6.2). The entire domain is quite homogeneous and of poor quality. The thickness of concrete to reinforce a north–south tunnel, for example, the thickness of reinforcements and the rock bolt size, can be the same almost everywhere. Figure 6.15c shows another reality: in almost the entire upper half of the domain, there are many more fractures than those predicted. On the contrary, the southern sector is of better quality. Thanks to this level of detail, it is possible to consider reducing part of the expense in the south to reinforce the structures in

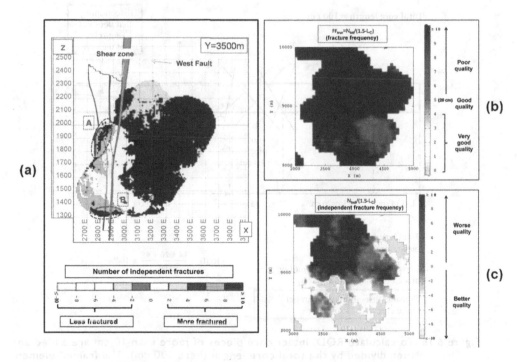

Figure 6.15 'Independent' fractures. (a) Vertical section of independent fractures in Chuquicamata. Domains A and B have many more fractures than they should when considering the directional concentration and crushing intensity. (b) Horizontal section of FF_{true} in Radomiro Tomic. (c) The same section showing the independent fractures. FF, fracture frequency.

the north, and crossing the boundary between these two sectors can be complicated. Figure 6.15b represents the quotient of two variables, a quotient that summarizes too much; Figure 6.15c is the joint utilization of the two variables of the previous quotient.

6.3 Rock quality designation and directionality

Like FF, RQD (Deere et al., 1967; Stagg and Zienkiewicz, 1968) aims at measuring the degree of rock fracturing by calculating the percentage of intact core pieces greater than 10 cm (Figure 6.16).

This additive variable could be mapped by kriging if it were not directional (that is, dependent on the direction of the drill hole) and subject, such as FF, to the sampling bias evidenced by Ruth Terzaghi in 1965: the measurement of the spacing between discontinuities, like their count, depends on the angle between the direction of the drill hole and that of the fractures. Suppose that, locally, the fractures are parallel planes with the same direction (Figure 6.17a) and consider two successive fractures (Figure 6.17b). While the spacing measured in direction 1 is l_1, it should be $l_1 \sin(\theta)$ to convert this spacing into an objective quantity independent of the drill-hole direction

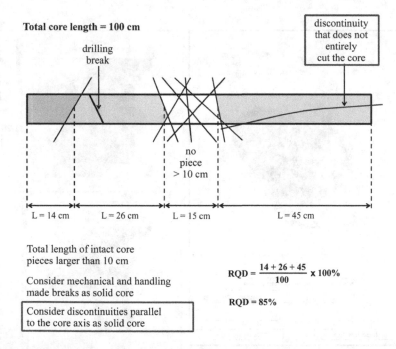

Total core length = 100 cm

drilling break

discontinuity that does not entirely cut the core

no piece > 10 cm

L = 14 cm   L = 26 cm   L = 15 cm   L = 45 cm

Total length of intact core pieces larger than 10 cm

Consider mechanical and handling made breaks as solid core

Consider discontinuities parallel to the core axis as solid core

$$RQD = \frac{14 + 26 + 45}{100} \times 100\%$$

$$RQD = 85\%$$

Figure 6.16 To calculate RQD, intact core pieces of more than 10 cm are added and then divided by the total core length (here 100 cm). The framed element shows another less common cause of dependence on the drill-hole direction: when a fracture cuts the core along its length, the element is still considered intact. It is sufficient then that the direction of the drill hole varies by a few degrees so that what could have been classified as intact is considered as fragmented. RQD, rock quality designation.

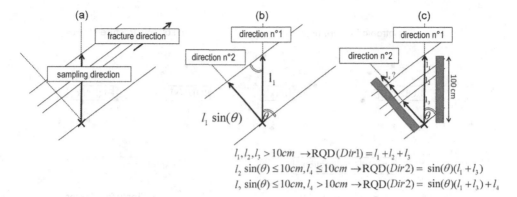

$$l_1, l_2, l_3 > 10cm \rightarrow \text{RQD}(Dir1) = l_1 + l_2 + l_3$$
$$l_2 \sin(\theta) \le 10cm, l_4 \le 10cm \rightarrow \text{RQD}(Dir2) = \sin(\theta)(l_1 + l_3)$$
$$l_2 \sin(\theta) \le 10cm, l_4 > 10cm \rightarrow \text{RQD}(Dir2) = \sin(\theta)(l_1 + l_3) + l_4$$

Figure 6.17 Directional bias of RQD. (a) Hypothesis of a local network of aniso-tropic fractures. (b) The intact distance between two fractures depends on the direction of the measurement according to the angle between this direction and the fracture. (c) While segment l_2 is incorporated into RQD when the direction of the drill hole is direction n°1, it ceases to be when this direction is perpendicular to that of the fractures. Therefore, a common factor $\sin(\theta)$ is not applicable to RQD to correct biases due to directionality. RQD, rock quality designation.

and perpendicular to the direction of the fractures. This problem also exists for FF. In the latter case, it is possible to apply a corrective factor $1/\sin(\theta)$ to each directional class to which the fracture belongs, which has been done previously for the directional concentration that considers corrected counts (Eq. 6.9). These corrections have been the subject of numerous developments to take into account, among other things, the fact that the definition of a 3D direction requires two angles. In addition, dividing by $\sin(\theta)$ tends to excessively increase the number of fractures when the angle θ tends to 0, hence the search for alternatives (Priest and Hudson, 1981; Chilès et al., 2008).

For RQD, such a correction, although imperfect, is impossible because it should not be applied to the constitutive sum of RQD, but to each of the contributing pieces, since, after individual correction, some pieces may appear below 10 cm and, therefore, be omitted from the sum. An example of this is the value $l_2 \sin(\theta)$ in Figure 6.17c. In this figure, the fact of working with a constant total core length means that a new piece is introduced in the calculation of RQD along direction 2, increasing the difference with the calculation along direction 1.

Assuming, for simplicity, that all the drill holes have the same direction and that the fractures at the scale of a kriging neighborhood are distributed in all the directions (Figure 6.18a), the directionality of RQD can be neglected, and its nondirectional map-ping then informs on the spatial variations of the degree of rock fracturing. Now, if the fractures locally follow preferential directions that change from one point x to another x' (Figure 6.18b), then the aforementioned argument does no longer hold, even if the drill-hole direction does not vary. The prediction $\text{RQD}^*(x)$, based on a set of samples around x, is no longer comparable with $\text{RQD}^*(x')$, based on other samples around x', because these two predictions differ by the samples used. A mapping based on these values represents not only the spatial variations of the rock fracturing but also the

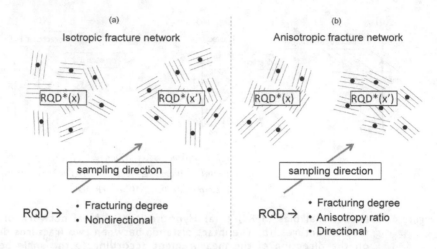

Figure 6.18 Fracture anisotropy and measurement direction. The points symbolize core samples. Does RQD measure the degree of fracturing of the rock? True in (a), false in (b) where a spatial variation of the anisotropy is indirectly measured since, locally, fractures tend to align along preferential directions, which does not happen in (a). RQD, rock quality designation.

spatial variations of the fracture anisotropy, which moves it away from the objective sought. As in the case of the mapping of FF_{true} that previously made the quotient of two regional quantities and, therefore, conveyed less information, the mapping of RQD is no longer informative by mixing two messages. Here, the deep nature of the concept of additivity is discussed; we will take the time to develop this point later.

The problem is further complicated when the drill-hole direction changes in space. A mapping of RQD by classes of sampling directions produces as many maps as there are directions (Figure 6.19a–f). A map that indiscriminately uses all the core samples only represents the most frequent drill-hole direction (Figure 6.19g). These maps illustrate the problem of the variations in the fracture orientations, since the impact on RQD is the same, either because the drill-hole direction changes or because the fracture direction varies from one point to another.

6.4 Rock quality designation and fracture frequency

How to correct RQD? Such a correction is necessary because both RQD and FF are components of the RMR attribute mentioned in the introduction of Section 6.2. If one is corrected, but not the other, there is a risk of destroying the link between these two variables, thus reducing the relative influence of one on the other. This problem is illustrated in the following scatter plots (Figure 6.20). A thick curve that represents a model defined by Stephan Priest and John Hudson in 1976 is shown, where RQD is expressed as a function of FF:

$$RQD = 100\, e^{-0.1FF}(0.1FF + 1). \tag{6.16}$$

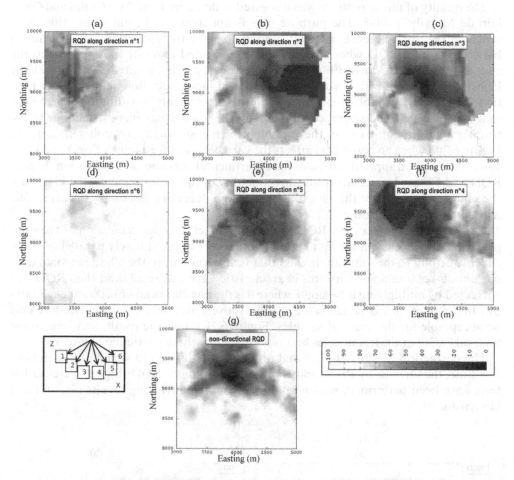

Figure 6.19 Directionality of RQD. In the Chuquicamata mine (and the surrounding mines), the drill holes are contained in east–west vertical planes for a better sampling of the variability transverse to the West Fault, so that a single angle is sufficient to determine the drill-hole direction. Six angular classes have been defined, and only the drill holes belonging to each class are used to calculate the maps (a) to (f). Map (g) indiscriminately uses all the drill holes. It resembles maps (c) and (e) because directions n°3 and 5 are majority. These maps illustrate the danger of looking at RQD without considering its directionality that depends in the same way on the direction of the fractures and that of the drill holes. RQD, rock quality designation.

This model assumes that the spacing between two consecutive discontinuities that intersect a drill hole follows a negative exponential law, which allows associating with each value of FF a distribution of spacing whose law is:

$$f(x) = FF\, e^{-FF\, x} \tag{6.17}$$

where $f(x)$ is the frequency of the spacing x.

The quality of this hypothesis was discussed in detail by Jean-Paul Chilès and Ghislain de Marsily in 1993. The purpose here is not to revisit it, but to use this theoretical curve as a reference, a visual support to show the distortions of correlations between FF and RQD when one variable is corrected and not the other, as in the case of Figure 6.20a where FF is corrected as suggested by Terzaghi. The formula by Priest and Hudson, which works well in general, does not allow modeling the relation between the variables, compared with the experimental conditional expectation that constitutes a good reference (thin curve). Having become inconsistent due to the transformation of only one of the variables, the measurements get coherent again in Figure 6.20b when no correction affects the variables. The formula by Priest and Hudson is then a good approximation of the experimental conditional expectation. Even if RQD and FF are biased due to the directionality of the measurements, the biases are the same, and the link of the variables is preserved. It is not satisfactory to correct one without correcting the other.

Can one still try to correct RQD? When the directional concentration of the sample is equal to 1, it is known that all the fractures are approximately parallel or, more exactly, belong to the same angle class that is also known. In the Chuquicamata data set, about 4,500 samples with this characteristic were extracted, and then RQD was corrected by multiplying it by $\sin(\theta)$, where θ is the angular (known) class of each sample. In doing so, the corrected value of RQD is overestimated, but this procedure may be acceptable for this class of samples. Figure 6.21 shows the result, focusing on the interval [0, 20] of the frequencies best represented by these particular samples.

The correction of RQD is poor and produces a reduction in variability such that, this time, the formula by Priest and Hudson overestimates RQD (Figure 6.21c). Other tests have been performed, without producing more convincing results (Séguret and Guajardo, 2015).

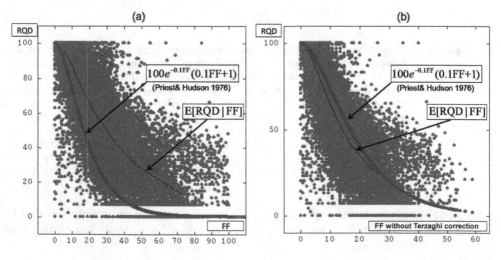

Figure 6.20 RQD and FF (1 of 2). Thick curve, model by Priest and Hudson; thin curve, conditional expectation of RQD knowing FF. (a) FF is corrected according to Terzaghi (b) FF is not corrected. FF, fracture frequency; RQD, rock quality designation.

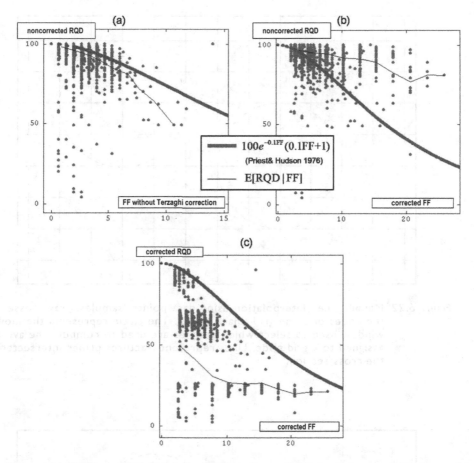

Figure 6.21 RQD and FF (2 of 2). Thick curve, model by Priest and Hudson; thin curve, conditional expectation of RQD knowing FF. (a) Neither FF nor RQD is corrected; (b) only FF is corrected; (c) both FF and RQD are corrected. FF, fracture frequency; RQD, rock quality designation.

6.5 Developments about directionality

FF and RQD are mainly used for two purposes:

- *Objective 1*: a direct mapping of rock quality domains. By mapping, we mean predicting values at the nodes of a fine grid by using the nearby samples. The values of FF and RQD produced on the grid represent the same volume of material as that of the samples, which are supposed to all have the same support (a few meters in length by some inches in diameter) (Figure 6.22).
- *Objective 2*: characterizing the rock quality at a block support using samples in and around the block. The calculation is associated with a block volume much larger than that of the samples, on the order of hundreds to thousands of cubic meters (Figure 6.23).

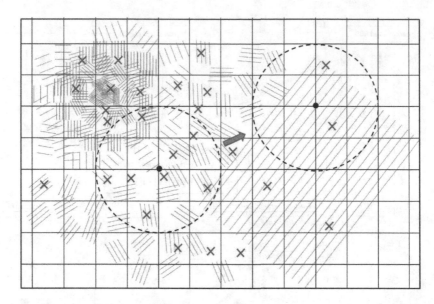

Figure 6.22 Mapping, i.e., interpolation using data points (samples, gray crosses) at the nodes of a fine grid (black point). The circle represents the moving window used to select which samples are used to compute the average assigned to a grid node. Lines represent fractures planes intersected by the cross-section.

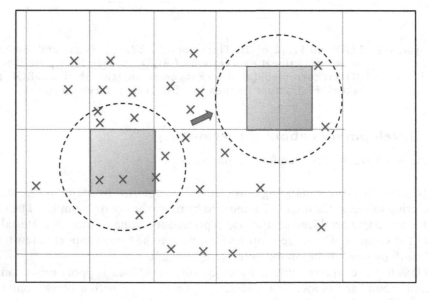

Figure 6.23 Calculating values on block volumes (gray) using samples around and inside the target block.

There are several issues associated with the pursuit of both objectives:

- the directionality of FF and RQD;
- FF and RQD are linear measurements (i.e., they are defined along a line) while one seeks to characterize the same quantity on a block volume;
- in some cases, FF and RQD are not directly used but are transformed into classes via a score scale (such as RMR), where values associated with the classification of FF, RQD and other variables are summed in order to quantify the rock quality. What is the impact on the spatial variability when the continuous FF and RQD values are transformed into discrete classes?

These issues have been well known for a long time, while robust workable solutions have been lacking. We propose here to clarify and develop the issues to support the wider use of FF and RQD, better knowing the pros and cons of such a use.

For simplicity, two cases are considered:

1. Case where there is no directional information associated with the fractures that intersect the core samples.
2. Case where there is directional information associated with the fractures that intersect the core samples.

6.5.1 No information on the direction of the observed fractures

Suppose, as shown in Figure 6.22, that the orientation and spacing of the fractures vary across the domain, but, at each sample location, the fractures are locally parallel and with a constant spacing. A more general situation, where, at any sample location, the fractures can have different orientations, will be considered in Section 6.5.2.

6.5.1.1 Measurement directionality

The measurement directionality relates to the situation where, at a given location x in the deposit, drillings made along different directions provide different values of the measured variable. An example is shown in Figure 6.24.

Two concepts are linked to directionality. The first one, called 'directionality bias', is the bias pointed out by Terzaghi (1965) who proposed a correction for FF, multiplying the observed FF value by the sine of the angle between the fracture and the sampling direction. As mentioned previously, this correction causes problems when the angle is too low and cannot be defined if the fracture is parallel to the drill-hole axis. Also, no correction is possible for RQD. Thus, in this case where no fracture orientation data is available, we do not propose a correction, so that a directionality bias of the measurements exists for both FF and RQD.

6.5.1.2 Fracture directionality

Correcting the measurement bias does not make the measurement nondirectional because the corrected quantity will still be linked to a particular fracture orientation: this is the second concept related to directionality. The underlying fractures (measured by

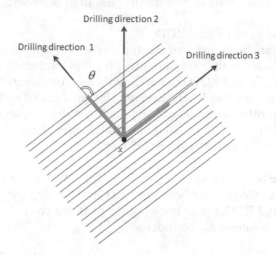

Figure 6.24 Example of directional bias when counting fractures (or when measuring their spacings for RQD), with a single fracture direction. Sampling direction 1 reveals 8 fractures; sampling direction 2 reveals 6 fractures; sampling direction 3 reveals 0 fracture (the sample diameter is constant). RQD, rock quality designation.

FF and/or RQD) are planar features that have a measurable orientation or direction, which may vary in space.

6.5.1.3 Why does directionality cause problems for mapping?

The aim of mapping is to show the spatial distribution of domains, or isopleths where FF or RQD has a constant value, the underlying idea being that, in these domains or along these isopleths, the rock quality is approximately the same because FF or RQD is the same. But such an interpretation is not possible without accounting for both the directionality bias of measurement and the fracture directionality.

We illustrate with an example in Figure 6.25. RQD is measured at four different locations by drilling along two directions in a domain where all the fractures have the same orientation. We define this case as *anisotropic fracturing* at the scale of the samples, because the fractures belong to a network with a preferential orientation.

The directionality bias of the RQD measurements breaks the true spatial continuity of RQD. We see that, when moving horizontally from left to right from x_1 to x_2, the RQD isopleths (built by the method shown in Figure 6.22) suggest an increase from a few percent to 100%, giving the map its bull's eyes. In this case of fracture anisotropy, where the fractures are oriented along the same direction and with the same spacing, RQD should be constant. However, the measurement bias changes from one point to another, and very different RQD values are predicted in space due to the interference of this measurement bias; it means that one cannot mix the sample values for mapping.

The correct way to proceed is to classify the samples according to their drill-hole direction (with some grouping) and mapping RQD by direction, producing as many maps as there are directional classes, as shown in Figure 6.26. The values of the samples

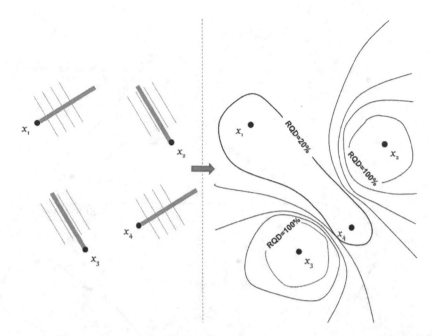

Figure 6.25 Two sampling directions in the case of anisotropic fracturing, i.e. all the fractures have the same orientation. The directionality bias of the RQD measurements breaks the spatial continuity of RQD, as represented by the isopleths (the sample diameter is constant). RQD, rock quality designation.

belonging to each directional class are comparable and compatible because they have the same measurement bias and can be used together for spatial interpolation purposes. However, we must not forget that the values are still biased in comparison with measurements that would be systematically perpendicular to the fracture plane.

6.5.1.4 Interpreting the directional maps

The resulting directional maps obtained after sample classification are not necessarily meaningful as they depend on another hypothesis: the isotropy or anisotropy of the fracture network itself. In our example in Figure 6.26, the fractures are parallel (anisotropic fracture network). Then, the maps made for each directional class represent the true spatial variation of FF or RQD, but still with a global bias that depends on the angle between the sampling direction and the fracture orientation.

The least biased map is derived from the samples that are closest to the perpendicular to the leading or dominant fracture plane. On the directional map showing the lowest RQD (or highest FF), changes in the orientation of the FF or RQD isopleths in space show the change in the fracture anisotropy direction, i.e., the change in the plane along which the fractures are aligned.

Our ability to interpret the spatial variations of FF and RQD depends on the isotropy hypothesis at the sample scale. As we suppose here in this subsection that there

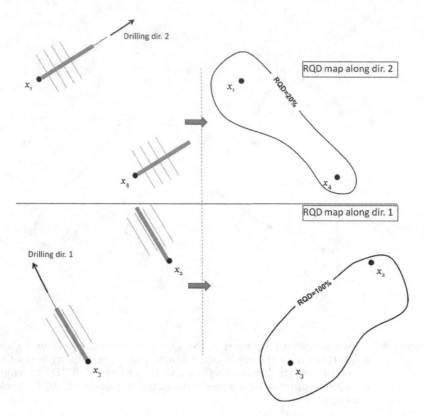

Figure 6.26 Example of mapping of RQD for each directional class of samples. (a) EW drill-hole direction; (b), NS drill-hole direction. RQD, rock quality designation.

is no information available about the orientations of the fractures observed on the samples, one can test the isotropy at the sample scale by using the directional maps as described earlier.

There are then two possibilities, detailed hereafter.

Case 1: The directional maps are similar

The fracturing can be considered isotropic at the sample scale because there are no preferential planar fracture orientations. It means that, for Objective 1 (mapping) and Objective 2 (change of support), one can proceed without further consideration of the sample directions because, inside each neighborhood used for prediction (as shown, for instance, in Figures 6.22 and 6.23), the measurement bias will be averaged in the same way for all the spatial predictions and will be comparable and compatible from one point to another. The map then gives a realistic image of the true spatial variability of FF and RQD.

In Figure 6.27, one sees an example of an isotropic fracture network where the fracture orientation changes quickly and where there is no preferential or dominant orientation. In this example, the fracture spacing (higher FF, lower RQD) is denser in the middle part of the area. The FF and RQD sample values are, on average, the same in any drill-hole direction.

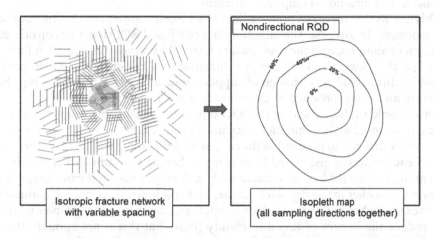

Figure 6.27 Case of an isotropic fracture network. Fractures exist in almost any orientation, and drill-hole samples are not directionally biased. The maps
produced by moving window averaging using the data from all the drill-
hole directions reflect the true spatial variability of RQD or FF, although
they may still be biased (Eq. 6.21). FF, fracture frequency; RQD, rock
quality designation.

When mapping with a moving average, the local measurement biases are always
averaged in the same way, and the map is a correct representation of the true spatial
variation, up to a global correction factor, as shown in the following (Eq. 6.21).

Even if the fracture network is isotropic, the sample values (and the averaged sample
values along an isopleth, for example) are still biased. It is proven that, when the angles
between core and fracture axes are uniformly distributed in the interval [0°, 90°], the
fracture count obtained when averaging 50 samples selected in a moving neighborhood gives 0.64×NB, NB being the true number of fractures measured perpendicularly to the fractures. The important difference between kriging and moving window
averaging is described in Section 6.5.1.6 hereafter.

An isotropic fracture network offers two practical advantages:

1. Values predicted on a grid using data with a directional sampling bias can be corrected by a multiplicative coefficient that removes the bias, as described earlier,
   and applied to the average values on the grid.
2. It is possible to produce a variable that becomes nondirectional, at any scale, including blocks of any size (change of support). By averaging linear FF or RQD
   measurements on a block support V, and because it involves all possible directions, we derive a quantity that can represent the average volumetric rock quality
   and size of intact material (one can call it RQD(V) and FF(V)). Such a change of
   support will then depend on other hypotheses such as, for example, the average
   length of the fractures or a local distribution of these lengths, parameters that are
   often used in geotechnics for simulating fracture networks. In the particular case
   of an isotropic fracture network, this can be done at a block support and conditioned by the parameters coming from the geostatistical analysis of the samples.

Case 2: The directional maps are different

This means that there is a preferential fracture orientation, i.e., the fracture network is anisotropic. In this case, the directionality of FF and RQD must be considered in all subsequent analyses, and the rock quality concept becomes directional, a linear (or at most 2D) characteristic, and never a volumetric one. For example, RMR predictions would be directional; any change of support must use only samples having the same direction and would produce block values that are also directional.

An example of directionality of RQD due to an anisotropic fracture network has been encountered at the Chuquicamata mine (Figure 6.19).

It is more difficult to proceed in the case of an anisotropic fracture network because, even if each map is constructed from samples having the same direction, there is no information about the bias associated with each map, a bias that can change from one domain to another in the deposit because, by hypothesis, there are local anisotropies, and these can change from one place to another. Classifying the samples by direction may reduce the impact of this uncertainty (bias), but this is not enough: the spatial change in the fracture anisotropy (fracture orientation) may produce a change in the bias, exactly in the same way as if the sampling direction had changed.

In Figure 6.28, we see an anisotropy associated with large domains and two sampling directions. A major fracture causes the anisotropies with the two orientations. The two directional maps are different, and one has to consider the drill-hole direction in all the subsequent analyses. Without any more information about the fracture

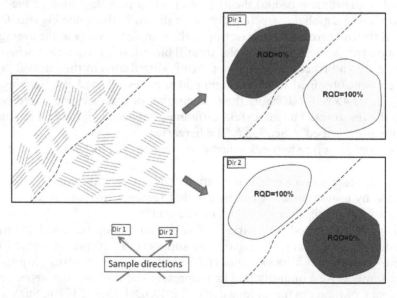

Figure 6.28 Anisotropy associated with large domains and two sampling directions. The dotted line shows a major fracture causing anisotropies with the orientations shown as gray lines. The directional RQD maps are completely different. In the absence of information about the fracture orientation, the importance of the bias associated with each map is unknown, making their interpretation problematic. RQD, rock quality designation.

orientation, the importance of the bias associated with each map is unknown, making difficult their interpretation.

The solution consists in searching for domains where FF is higher (or RQD lower) when comparing the directional maps, as in Figure 6.28. This is a way to associate the dominant fracture orientation with each domain.

6.5.1.5 What can be done, then? There is a need for just one map, not a set of directional ones!

When maps only differ due to the sampling directional bias, comparing them at given locations can lead to finding 3D domains where FF is higher (or RQD lower), in which case the sampling direction is almost perpendicular to the main fracturing anisotropy. In such domains and for this direction, one can consider that there is no sampling bias. Then, instead of dealing with a set of directional maps and RMR tables throughout the analysis, one could choose to use the data and map associated with the sampling along the least biased direction (Figure 6.29).

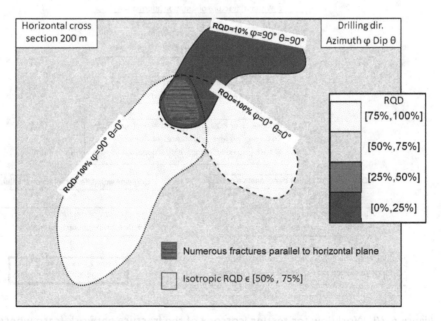

Figure 6.29 A map that captures the main results of an analysis of directional maps. The light gray region is where the fracture network is isotropic; here the directional RQD maps are similar and RQD values are in the interval 50%–75%. The white region shows directional domains where RQD values are high (above 75%) when using samples from horizontal drill holes in any azimuth orientation. The region shown in dark gray exhibits very low RQD when obtained from vertical drill holes; however, the fact that RQD values are low only in vertical drilling means that inside the intersection zone, the fractures are horizontal. RQD, rock quality designation.

Again, we are dealing here with the case where no fracture orientation data is available. The workflows are shown in Figures 6.30 and 6.31.

6.5.1.6 What kind of predictor should be used when the data are biased?

Predicting FF and RQD values should not be made by ordinary kriging (OK) because OK produces a linear combination of the measurements with weights that depend on the distance between the sample and the target point. Let us show why now.

Firstly, it is equivalent to consider fractures with changing angles and a unidirectional set of fractures with changing sampling directions. In the following proof, the fracture spacing and the core diameter are considered to be constant.

The geometry of the fracture orientation relative to three drill-hole directions is shown in Figure 6.24. When measured perpendicularly to the fracture plane, the number of fractures is denoted as NB, a value that becomes NB sin(θ) when the sample core axis is not perpendicular to the fracture plane.

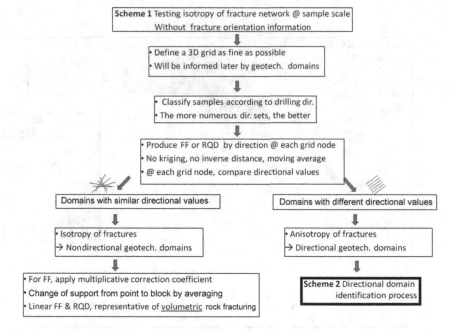

Figure 6.30 Workflow for testing isotropy of the fracture network (case where there is no measurement of fracture orientation). The work starts with a classification of the samples according to the drill-hole direction. The similarity of maps produced per directional class at the nodes of a grid is the basis for determining the directionality of the measurements: if maps are similar between the directions, the fractures are isotropic and geotechnical domains are nondirectional. After corrections for directional bias, all proceeding work can ignore the drill-hole direction as the variables are additive and a change of support is valid. If the maps are not similar between the directional classes, the workflow must proceed as per Scheme 2 shown in Figure 6.31. FF, fracture frequency.

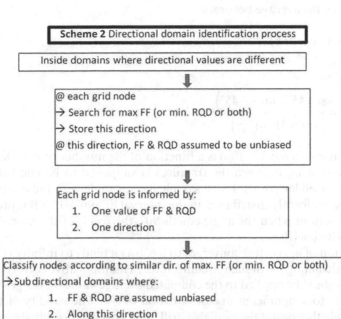

Figure 6.31 Workflow when directional maps are different (case where there is no measurement of fracture orientation). All work is per directional class. The first task is to find the drill-hole direction that gives the least biased measurements and use it to produce values of FF and RQD on grid nodes by moving window averaging. Along this direction, the FF and RQD predictions are unbiased. The grid nodes can be classified and grouped to produce regions of similar FF and RQD values. FF, fracture frequency; RQD, rock quality designation.

Suppose that we have $n+1$ samples associated with angles θ regularly distributed in the interval $[0°, 90°]$ and we average these measurements to produce a prediction:

$$NB^* = \frac{1}{n+1}\sum_{k=0}^{n} NB\sin\left(k\frac{90°}{n}\right) = \frac{NB}{n+1}\sum_{k=0}^{n}\sin\left(k\frac{90°}{n}\right). \tag{6.18}$$

To calculate the sum, we refer to Adams and Hippisley (1922):

$$\sum_{k=0}^{n}\sin(x+ky) = \frac{\sin\left(x+\frac{ny}{2}\right)\sin\left(\frac{n+1}{2}y\right)}{\sin\left(\frac{y}{2}\right)}. \tag{6.19}$$

By setting $x=0$ and $y=\frac{90°}{n}$, we obtain:

$$\sum_{k=0}^{n}\sin\left(k\frac{90°}{n}\right) = \frac{\sin(45°)\sin\left(\frac{n+1}{n}45°\right)}{\sin\left(\frac{45°}{n}\right)}. \tag{6.20}$$

The result of the average becomes:

$$NB^* = NB \times \text{coeff}(n). \tag{6.21}$$

with

$$\text{coeff}(n) = \frac{\sin(45°)\sin\left(\frac{n+1}{n}45°\right)}{(n+1)\sin\left(\frac{45°}{n}\right)}. \tag{6.22}$$

The reduction coefficient coeff(n) is a function of the number of samples n used in the average. The spacing between the fractures is supposed to be constant, the sample support has a fixed length, and each sample has an angle with the fracture so that the n samples are uniformly distributed in the interval [0°, 90°]. If NB represents the true number of fractures when the angle equals 90°, the result of the average reduces this number to NB×coeff(n).

As a function of n, coeff(n) converges to 0.636 as n tends to infinity (Figure 6.32).

So when calculating FF in such circumstances, a multiplicative correction factor equal to 1/coeff(n) should be applied to the computed average value on the grid, as the bias in the data leads to a significant underestimation of FF. The validity of this correction depends on whether or not the available drill holes cover uniformly the angle interval.

In contrast, the average should not be predicted by OK because OK produces a linear combination of the measurements with weights λ_k that depend on the distance between the sample and the target point. Then, prediction (6.18) is replaced by:

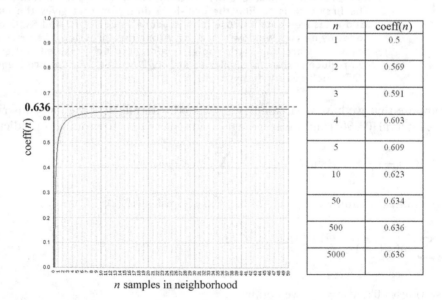

n	coeff(n)
1	0.5
2	0.569
3	0.591
4	0.603
5	0.609
10	0.623
50	0.634
500	0.636
5000	0.636

n samples in neighborhood

Figure 6.32 Correction coefficient coeff(n) for FF as a function of the number *n* of samples used to compute the average. NB represents the true number of fractures when the angle equals 90°. The moving window averaging reduces this to NB × coeff(n). FF, fracture frequency.

$$\mathrm{NB}^{\mathrm{OK}} = \mathrm{NB} \sum_{k=0}^{n} \lambda_k \sin\left(k\frac{90°}{n}\right) \tag{6.23}$$

with

$$\sum_{k=0}^{n} \lambda_k = 1. \tag{6.24}$$

In Figure 6.33, we see a particular sampling situation where a data point x_0 is closer to the target in comparison with the other points. Accordingly, the kriging weight assigned to this point is large (depending on the variogram model), let us say 0.8. Due to constraint (6.24), the n remaining points will receive a weight approximately equal to $0.2/n$ (we do here quick hand calculations to give an idea on orders of magnitude), and Eq. (6.23) becomes:

$$\mathrm{NB}^{\mathrm{OK}} = \mathrm{NB}\left(0.8\sin\left(i\frac{90°}{n}\right) + \frac{0.2}{n}\sum_{k\neq i}\sin\left(k\frac{90°}{n}\right)\right). \tag{6.25}$$

For example, if $i=0$ and $n=50$, $\mathrm{NB}^{\mathrm{OK}} \approx 0.129$ NB.

If the data were not biased, then this property of kriging would be beneficial, as the closer the sample to the target, the higher correlation it has with the target value

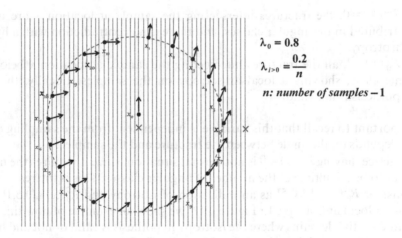

$$\lambda_0 = 0.8$$
$$\lambda_{i>0} = \frac{0.2}{n}$$

n: number of samples − 1

Figure 6.33 Sampling situation where data point (x_0) is closer than other data to the target point (gray cross). This point gets a kriging weight of 0.8, significantly higher than any other point x_1 … x_{21}, which receive similar weights as they are located at the same distance from the target. FF data are biased according to the angle between the fracture and the sampling direction. A large kriging weight applied to samples with large bias will produce a strongly biased prediction. In this case, the predicted value is much lower than the true FF measured perpendicularly to the fractures. At another target, the bias may change, and the different predictions cannot be compared. FF, fracture frequency.

(as measured by the variogram). But in this case of biased data, one has fracture count or RQD data from several sampling directions, which is formally equivalent to having a single sampling direction with several fracture orientations. The data are biased depending on the angle between the fracture and the sampling directions. The large kriging weight magnifies the importance of the bias and produces a much lower (less than 20%) prediction of the true FF or RQD measured perpendicularly to the fractures, and less than a moving average prediction (0.636 times the true measurement).

When a specific data point receives a large kriging weight due to its closeness to the target point, and when the data value at that point is severely biased, the OK prediction of FF is now close to 10% of the true value. This percentage could change quickly at short distances between target points.

By assigning a different weight to each data point, kriging 'disarranges' the sampling directional homogeneity and produces a spatial bias, making the final result useless. Consequently, kriging (as well as inverse distance weighting) is not recommended to predict FF, a moving average is globally less biased because it avoids assigning high weights to strongly biased samples. The calculation for RQD is more complex, but the recommendation is the same: use a moving average with enough samples, say at least 30, instead of kriging.

6.5.2 With information on the direction of the observed fractures

$R_\theta^2(x)$ as defined by Eq. (6.12) can be used to classify the samples. We recall that:

- If $R_\theta^2(x) = 0$, the fractures intersecting the sample at location x are uniformly distributed in the angular classes, and there is no contradiction with a hypothesis of isotropy.
- If $R_\theta^2(x) = 1$, all the fractures intersecting the sample at location x belong to the same class, showing a local anisotropy on the samples. Locally, the isotropy hypothesis is not valid.

It is important to recall that this variable is independent from the sampling direction; it only depends on the angle between the fracture and the sample.

In practice, having $R_\theta^2(x) = 0$ is not a usual situation, since it requires the number of fractures to be a multiple of the number of angular classes. This is the reason why one can consider $R_\theta^2(x) \in [0, 0.5]$ as an indication of isotropy and $R_\theta^2(x) \in]0.5, 1]$ as an indication of local anisotropy. In Figure 6.12, we show an example of how this attribute can define spatial domains where the isotropy hypothesis is valid or not and how it can be linked to structural domains defined by major faults.

Séguret (2016) also shows a link between the directional concentration and the number of fracture sets, which is frequently available in geotechnical data sets and may be an alternative to the directional concentration for detecting local anisotropies. Having $R_\theta^2(x)$ makes it possible to modify the workflow, as shown in Figures 6.34 and 6.35.

The modeling of domains where geotechnical variables are directional can use again the directional information. As the sampling direction is supposed to be known, it becomes possible to know which classes contain more fractures, and their average absolute orientation (azimuth and dip).

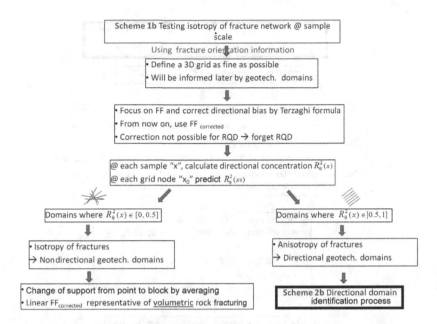

Figure 6.34 Modified workflow: testing isotropy with available fracture orientation data. FF data are Terzaghi corrected. At each sample location, directional concentration values are computed and used to predict the directional concentration at each grid node. The grid is divided into regions defined by the predicted directional concentrations: within the interval [0, 0.5], the fractures are considered isotropic, and FF is additive without directionality. At the nodes where the directional concentration belongs to the interval]0.5, 1], the fractures are locally anisotropic, and all preceding analyses must consider the sampling direction, as shown in Figure 6.35. FF, fracture frequency.

When the number of angular classes is n_θ and the number of drilling directions is n_{dir}, this gives a maximum $n_\theta\,n_{dir}$ preferential directions with which one can associate a number of fractures. This leads to as many directional maps as there are such directions. For each map, the samples used are the ones that have the same preferential direction. The attribute mapped in this way is the number of fractures in each direction, and all subsequent works must be based on the number of fractures and not any more on FF or RQD.

6.6 Spatioangular geostatistics

The main lesson one can retain is that the rock quality classification must account for directionality because the isotropy hypothesis at any scale is not always true.

The previous work tried to characterize such a directionality, depending on whether some information on the fracture orientations is known (Section 6.5.2) or not (Section 6.5.1). We can at least say that the workflows are laborious. Recent developments can, however, simplify a lot these workflows, by an astute definition of the space in which the measurements are regionalized.

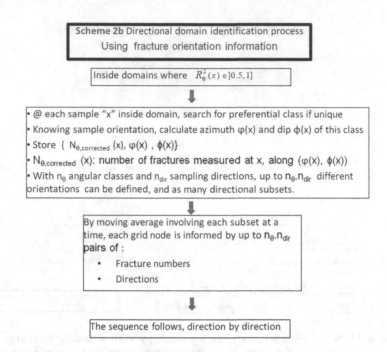

Figure 6.35 Modified workflow for grid domains where the predicted directional concentration belongs to the interval]0.5, 1].

Since a measurement varies according to both its position and orientation, the idea is to assign it a vector of coordinates (x) in the geographical 3D space and a vector (u) on the 2D half-sphere. All in all, one obtains a regionalization in a 5D space corresponding to the product on the 3D Euclidean space and the 2D sphere, so that each measurement is now indexed by its easting, northing, elevation, azimuth and dip. Instead of making predictions and simulations conditioned to a particular direction, this new paradigm allows FF or RQD to be interpolated at any place in the geographic space, for any direction. This is an important simplification because the concept of directional bias disappears because, for each location x, values are produced for any direction u, as shown schematically in Figure 6.36.

Let us denote by $Z(x, u)$ the random field associated with the regionalized variable under study (FF, RQD or another geotechnical property), with x the vector of geographic coordinates and u the vector of orientation on the half-sphere. Some simplifying hypotheses are needed to infer the spatial correlation of such a random field. Concerning the variations in the 3D geographical space, one commonly assumes second-order stationarity, so that the covariance or the variogram between two variables located at x and x', respectively, only depends on the geographical separation $h = x' - x$. Concerning the variations on the 2D sphere, second-order stationarity is no longer applicable, but one can trade off the invariance of the first- and second-order moments under a translation against an invariance under a rotation, which corresponds to an assumption of isotropy. In such a case, the first-order moment (expectation) is constant on the sphere, and the second-order moment (covariance) between two variables

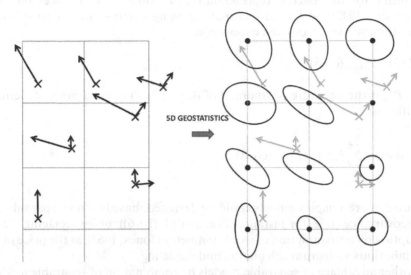

Figure 6.36 Interpolation of a variable regionalized in a 5D space. For sake of simplicity, the representation is made in a 2D plane, but a spheroid, possibly multi-modal, should be represented, which characterizes the directional variations at a given location *x* of the geographic space. Notice that it is possible, at each location *x*, that a sample has several directional values. This is the case when fractures are classified according to directional classes. Predictions or simulations (tens or hundreds) can be done in this 5D space.

located at u and u' only depends on the angular separation $\delta = \arccos(<u', u>)$, also known as the geodesic distance between u and u'. Such a hypothesis is plausible, for instance when the fracture network is isotropic or when it is locally anisotropic with anisotropy directions that vary in space and are, globally, uniformly distributed on the sphere.

For calculating experimental covariances/variograms under the previous hypotheses, it is sufficient to classify the data pairs according to their geographic separation (h) and angular separation (δ), all other things being equal. The modeling can be done by a linear model of regionalization, consisting of a sum of basic nested structures. One can write the theoretical covariance or variogram as:

$$C(h,\delta) = \sum_{s=1}^{S} b_s\, C_s(h,\delta) \tag{6.26}$$

$$\gamma(h,\delta) = \sum_{s=1}^{S} b_s\, \gamma_s(h,\delta) \tag{6.27}$$

with S a positive integer and $b_1, \ldots, b_S$ are positive scalar coefficients. As for the choice of the basic structures, Sánchez et al. (2019) propose separable covariance models of the form $C_s(h,\delta) = \rho_s(h)\,\rho_s(\delta)$, where $\rho_s(h)$ is a stationary geographic covariance (e.g., a spherical or an exponential model), while $\rho_s(\delta)$ is an isotropic angular covariance.

Accounting for the spectral representation of isotropic covariances on the sphere (Schoenberg, 1942) and for the fact that changing u into $-u$ or u' into $-u'$ should not modify the covariance, one can choose ρ'_s as:

$$\rho'_s(\delta) = P_{2n(s)}(\cos\delta) \tag{6.28}$$

where P_{2n} is the Legendre polynomial of degree $2n$. The covariance structure finally simplifies as

$$C(h,\delta) = \sum_{s=1}^{S} b_s\, \rho_s(h)\, P_{2n(s)}(\cos\delta) \tag{6.29}$$

Of course, more complex models could be designed, based on nonseparable spatioangular covariances (see, for instance, Porcu et al. (2016)), or by replacing the isotropy assumption by an assumption of axial symmetry (Jones, 1963), at the price of a (much) more laborious variogram calculation and modeling.

Another advantage of separable models or combination of separable models of the form (6.29) is the possibility to adapt the spectral and turning bands algorithms in order to simulate the random field $Z(x, u)$ at any position x and any direction u. Along each turning line, it suffices to (1) sort an integer s in $[1,S]$ with a probability proportional to b_s, (2) simulate a zero-mean random field T_s in the geographic space with covariance $\rho_s(h)$, (3) simulate a zero-mean random field W_s on the sphere with covariance $P_{2n}(\cos\delta)$ and (4) generate the simulated random field along the turning line as the product of the geographic and angular components T_s and W_s, up to a normation factor (Sánchez et al., 2019). The simulation of the random field W_s can be done by considering spherical harmonic functions (Emery and Porcu, 2019) or Gegenbauer waves (Alegría et al., 2020). This workflow is summarized in Figure 6.37.

An interesting application to the modeling of the weak vein intensity in the El Teniente deposit is presented by Sánchez et al. (2019), who use the same data as that introduced in Section 6.1 of this chapter, but focus on the linear frequency P_{10} instead of P_{32}. As the study is oriented to simulation, the P_{10} data are first normal-score transformed, and their experimental variogram is calculated by accounting not only for the geographic separation between the data pairs (distinguishing between the horizontal and vertical directions, recognized as the main directions of continuity in the geographic space) but also for their angular separation (Figure 6.38). Declustering weights are assigned to the data in both the normal score transformation and the experimental variogram calculation, in order to account for the irregularities in the sampling design. One observes that the behavior of the experimental variogram for short geographic separation distances (0–100 m) depends on the angular separation between the data, as it increases for small angular separations ($\leq 30°$) but decreases for larger angular separations. To model this behavior, a separable model is considered, which consists of the product of a short-range spherical covariance (40 m in the horizontal plane and 60 m in the vertical) and a Legendre polynomial of degree 2, which switches from positive to negative values when the angular separation goes from 0° to 90°. A second spherical-Legendre model with larger ranges in the geographic space (500 m) is also introduced, as well as two other spherical structures that only vary with the

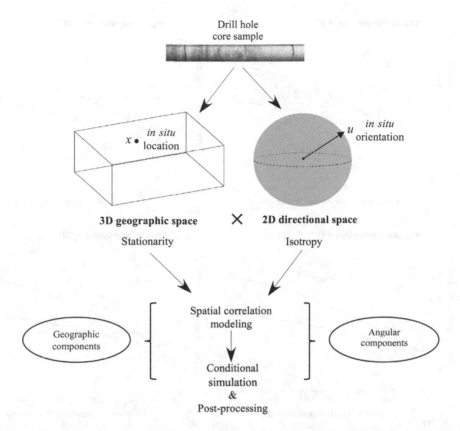

Figure 6.37 Workflow for the 5D modeling and simulation of directional regionalized variables. Each core sample is assigned a vector *x* of geographic coordinates in the 3D space, and a vector *u* of orientation on the 2D sphere. Under an assumption of stationarity in the geographic space and isotropy on the sphere, both the fitting of the covariance or variogram and the simulation of the variable can be performed by considering products of geographic and angular components.

geographic separation (their angular component corresponding to a Legendre polynomial of degree 0, which is identically equal to 1).

One hundred conditional simulations are constructed with the turning bands algorithm and averaged, in order to predict the weak vein frequency P_{10} at any spatial location and any direction on the sphere. This gives rise to two kinds of representations, which are helpful to structural geologists and geotechnicians to assess the geographic and directional regionalization of the weak vein frequency:

- geographical maps that show the spatial variations of P_{10} for a given direction *u* (Figure 6.39a);
- orthographic maps that show the directional variations of P_{10} for one or more geographical positions (Figure 6.39b).

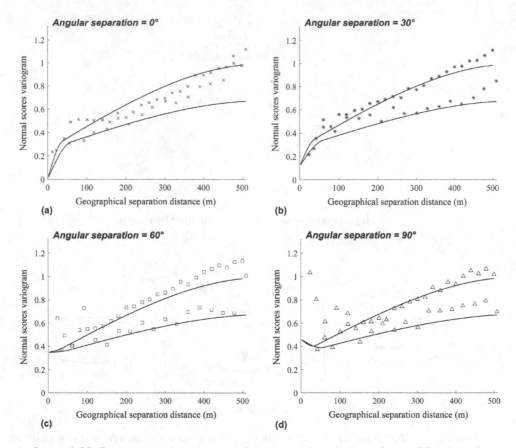

Figure 6.38 Experimental variogram of the normal score transform of P_{10} data along the horizontal and vertical directions (crosses, stars, squares and triangles), for geographic separation distances between 0 and 500 m and angular separations of 0° (a), 30° (b), 60° (c) and 90° (d) (tolerance ± 15°). The modeled variogram based on a combination of separable spherical-Legendre structures is superimposed (continuous lines).

It becomes unnecessary to remove any bias, and local anisotropies can be derived from analyzing the results in the 5D space, which is really the heart of the problem. The model is directional, as is the rock quality classification, throughout the analysis. For example, the weak veins turn out to be preferentially oriented in north–south vertical planes near the locations with coordinates (1,425 m, 400 m, 200 m) and (1,425 m, 300 m, 2,000 m), for which the vein frequency is the highest when it is calculated along the east direction, but the contrary happens near locations with coordinates (1,425 m, 350 m, 200 m) and (1,325 m, 300 m, 2,000 m) (Figure 6.39b).

There is no doubt that the 5D approach represents the ultimate way of solving many of the problems encountered in geotechnics. It is currently at a research stage and not ready for industry use.

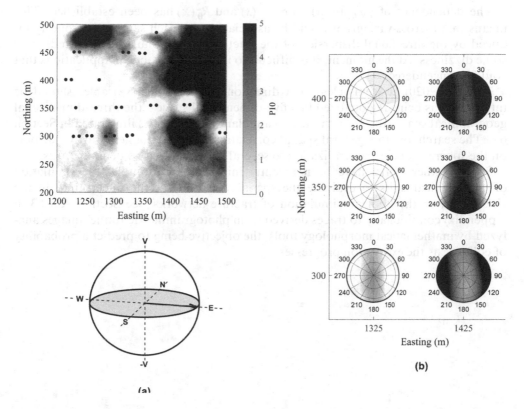

Figure 6.39 Average of 100 conditional simulations of the weak vein frequency. (a) Geographic map of the frequency calculated along the east–west direction for an elevation of 2,000 m; the positions of the conditioning data are superimposed. (b) Orthographic maps at six geographic positions with elevation 2,000 m, showing the expected directional variations of the weak vein frequency (Sánchez et al., 2019).

6.7 Conclusions

With respect to the corrections of FF and RQD and their link, the problem remains open and an investigation will be necessary where both variables, as well as the others involved in RMR, should be analyzed and modeled together. Other clues have been left unexplored. In particular, it has been found that the cross-variograms between the number of fractures and the fragmented core length are proportional to the regional correlation coefficient as defined in Eq. (6.13), which means that their statistical link is, in fact, geostatistical; one would therefore have, for the cross-variogram between the fragmented core length and the number of fractures:

$$\gamma_{N,LC}(h,x) = \sigma^2(x)\,\rho_{N,LC}(x)\gamma(h) = \sigma^2(x)\,f(R_6^2(x))\gamma(h), \tag{6.30}$$

where $\gamma(h)$ is a variogram with a unit sill.

The dependence of $\gamma_{N,LC}(h, x)$ on $\rho_{N,LC}(x)$ and $R_\theta^2(x)$ has been established. This means that a cross-variogram would be associated with each sample, a property induced by the directional dispersion of the fractures that, it must be admitted, gives some dizziness. At the moment, it is difficult to foresee the practical applications that could be engendered.

To date (2020), it seems clear that directional geotechnical variables should be interpreted as being regionalized in a five-dimensional space – the three-dimensional geographical space crossed with the two angular dimensions, as illustrated in Section 6.6. The search for more general spatial correlation models – not necessarily isotropic on the sphere – and the generalization to several variables are two pending tasks.

Several other works have been recently initiated, for example, the simulation of geotechnical units to evaluate the spatial variations of the uncertainty in the risk factor, or the Boolean simulation of fracture networks (see Section A.5.2.3 in Appendix) conditionally to traces derived from photogrammetric tunnel images analyzed by mathematical morphology tools, the objective being to predict a probability of collapse as the excavation progresses.

Chapter 7

Recovery and geometallurgy

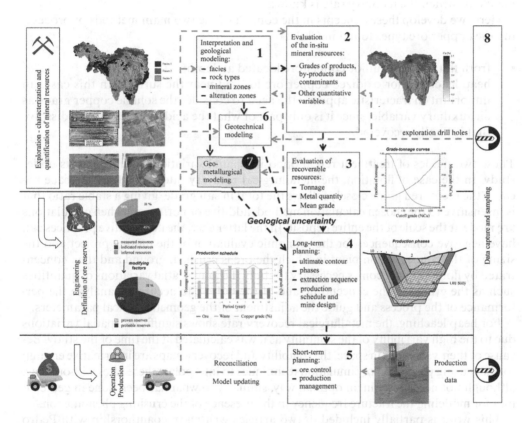

Until now, the total in situ copper grade has been studied and predicted in the deposit by taking into account, when possible and useful, the information on geological facies that control the grade variability. But it is the metal quantity that will actually be recovered after processing that really matters. This quantity depends on the metallurgical recovery rate, quotient of the recovered metal quantity and the original total metal quantity, which turns out to be the variable preferred by the metallurgists and geologists responsible for the economic evaluation of the deposit to reflect the performance of the process.

This recovery rate is a false friend: firstly, it could vary little in space and, therefore, not be very informative. Secondly, when production begins, the first results of the actual recovery may come with some surprises and not be up to the predictions: the recovery performance fluctuates much more than expected for several reasons later described. Also, the metallurgical recovery is a false friend because it is not additive, so it cannot be averaged arithmetically to represent the expected recovery of a block or a production volume.

These difficulties have led to the introduction of the concept of 'recoverable grade', the simple product of the recovery rate by the total in situ grade. By focusing on this service variable, the possibility of cokriging it with the total grade appears. It becomes a convenient way to compensate for the typically small number of laboratory measurements for which the recovery rate is known.

Here, we develop these concepts in the context of the two main methods of processing the copper ore types found in Chile:

- froth flotation for sulfide copper ore located at depth;
- heap leaching, for oxidized copper ore located near the surface. In this case, an important characteristic appears: the need to consider the soluble copper grade as an auxiliary variable, since it is only part of what the acid has been able to dissolve that is finally recovered.

These two modes of treatment reveal opposite spatial variations in the deposits under study. In the case of flotation, the metallurgical recovery rate varies little in space because the recovered grade closely follows the total in situ grade, giving a stable ratio that is insensitive to the chosen calculation method, additive or not, except when calculations are made at the scale of the entire deposit. In the latter case, the nonadditive practices can have negative consequences for the economic evaluation of the mining project and the sizing of the concentration plant in which the ore is crushed, ground and then concentrated by flotation. It becomes necessary to study the joint spatial variations of quantities such as the grades of the concentrate and of the tailing, to accurately analyze the performance of the process and guide the adjustment of the geometallurgical parameters.

For heap leaching, the metallurgical recovery rate shows significant spatial variations due to the high variability of the solubility as it was calculated at the time of the study. Because of their variations in space, the solubility and recovery maps are informative enough to allow using prediction or simulation techniques to interpolate these rates without bias, although not necessarily in an optimal way, similarly to what has been done in geotechnics for modeling the fracture frequency in the presence of the crushing phenomenon.

This work is partially included in two articles written in coauthorship with Pedro Carrasco in 2004 and 2008 (Emery et al., 2004; Carrasco et al., 2008).

7.1 Sulfide copper ore

7.1.1 Concentration by flotation

After grinding the ore into fine particles smaller than 250 μm, the resulting ore pulp is treated in flotation cells containing bubbles of gas (normally, air) and chemical reagents. These reagents act on the surface of the minerals to be recovered so that they become hydrophobic and rise in the cell without agglomerating, carried by bubbles and released from the gangue (Wills and Finch, 2015). On the surface of the cells, froth is recovered, and after several stages of treatment, a pulp with a high metallic content called 'concentrate' is obtained (Figure 7.1). The 'tailing' in the lower part of the cells is accumulated in huge artificial deposits, and could be processed again in the future if the copper price increases sufficiently. This process, widely used in the mining industry, allows moving from an average feed copper grade Z_H of the order of 1%, to more than 30% for the concentrate grade Z_C, depending on the mineral species carrying the copper, and less than 0.1% for the tailing grade Z_T. The subscripts here correspond to H for head (feed or total in situ grade), C for concentrate and T for tailing.

Before mine production begins, laboratory experiments are performed in order to forecast the flotation performance of the ore that will be processed. Samples are taken from the study area and processed in mini flotation cells that are small copies of the device to be used. Figure 7.2 shows an example of it. It is for this situation, prior to production and at a laboratory scale, that the study presented in the following is carried out.

7.1.2 Formalization

Two types of variables are involved to forecast the performance of the process:

- the tonnages T_H, T_C and T_T involved in the calculation of the feed grade Z_H, concentrate grade Z_C and tailing grade Z_T;
- the metal quantities Q_H, Q_C and Q_T associated with these grades.

These variables are related to each other, but not systematically. One can imagine a low tonnage of concentrate with a high metal quantity and vice versa. The spatial variations of these variables depend on both the spatial variations of the mineralogy and the response of the chemical processes to this mineralogy. To account for this typology of the variables, the following rates r and R are introduced:

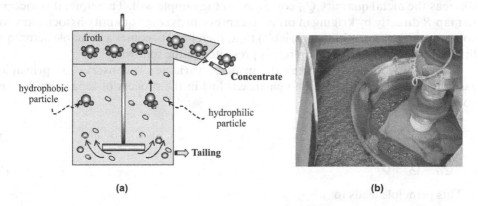

(a) **(b)**

Figure 7.1 Flotation cell. (a) Schematic representation; (b) froth concentrate on the surface of the cell. (Credit: Aldo Casali and Gonzalo Montes-Atenas.)

Figure 7.2 Flotation cell used in laboratory. (Credit: Gonzalo Montes-Atenas.)

$$r = \frac{T_C}{T_H} \tag{7.1}$$

$$R = \frac{Q_C}{Q_H}. \tag{7.2}$$

r is the mass recovery rate, whereas R is the metallurgical recovery rate, which is of highest interest to economic geologists. Although each of these two rates measures a particular aspect of the flotation performance, they do not have the same statute, mathematically speaking. Under the assumption that the density of the ore varies little in space (which is true for copper ore, due, among other things, to its low grades) and provided that, in each kriging neighborhood, only measurements made on the same support are used together, the tonnage T_H will not change from one sample to another, whereas the metal quantity Q_H contained in the sample will. Therefore, it is incorrect to map R directly by kriging in order to express an average quantity associated with a volume, unlike r (an additive variable) that, therefore, becomes a valuable actor, quite little known and used. Its main role is presented in the following.

To formalize the link between the involved variables, a conservation principle is assumed, according to which no particle is lost in the process of separation between concentrate and tailing:

$$T_H = T_C + T_T \tag{7.3}$$

$$Q_H = Q_C + Q_T. \tag{7.4}$$

This principle leads to:

$$Z_H = r Z_C + (1-r) Z_T \tag{7.5}$$

with

$$Z_C = \frac{Q_C}{T_C} \tag{7.6}$$

$$Z_T = \frac{Q_T}{T_T}. \tag{7.7}$$

In formula (7.5), r separates what is recovered from what is lost. This formula also indicates that it is sufficient to know Z_H, r and, indifferently, Z_C or Z_T, to express all the variability of the problem, which could lead to consider a cokriging of Z_H, r, Z_C or Z_T to predict each of these variables in an optimal manner. But this is not possible because Z_C and Z_T are not additive, since these grades refer to tonnages T_C and T_T that change from one sample to another.

If one remembers that:

$$r Z_C = \frac{Q_C}{T_H}, \tag{7.8}$$

formula (7.5) becomes:

$$Z_H = \frac{Q_C}{T_H} + \frac{Q_T}{T_H}, \tag{7.9}$$

the obvious decomposition of the feed copper grade into *partial grades*, a concept already found (several pages above and ten years later!) in the problem of constructing a block model. Here, the metal quantities are no longer distinguished based on the indicators of geological facies, but based on the indicators of the mineral particles found in the concentrate or in the tailing. Formula (7.9) is then interpreted as the additive conversion of the quantities Z_C and Z_T, which are not additive, into grades Z_R and Z_{TT}, which are additive, with:

$$Z_R = R Z_H = r Z_C = \frac{Q_C}{T_H} \tag{7.10}$$

$$Z_{TT} = (1 - R) Z_H = (1 - r) Z_C = Z_H - Z_R = \frac{Q_T}{T_H}. \tag{7.11}$$

The conclusion that is reached is to perform a cokriging of the additive variables Z_H, r, Z_R or Z_{TT} to optimally predict any of these variables.

7.1.3 Usual practices

At the time of the study, the laboratory technicians were measuring Z_H, Z_T and Z_C and calculating R using the following exact formula, which comes from the conservation principle:

$$R = \frac{Z_C}{Z_H} r = \frac{Z_C}{Z_H} \frac{Z_H - Z_T}{Z_C - Z_T}. \tag{7.12}$$

Calculated on each sample (index i), these ratios were averaged over the n available samples, to obtain a quantity $\tilde{R}$:

$$\tilde{R} = \frac{1}{n} \sum_{i=1}^{n} R_i. \tag{7.13}$$

Considered representative of the entire deposit or of a particular geological unit (facies), $\tilde{R}$ was then combined with the kriged grade $Z_H^K(V)$ of the block model to obtain a prediction of the recovered grade $Z_R(V)$ of block V:

$$\tilde{Z}_R(V) = \tilde{R} \, Z_H^K(V). \tag{7.14}$$

This practice raises several questions:

- formula (7.13) averages nonadditive quantities, the correct formula being

$$\hat{R} = \frac{\sum_{i=1}^{n} Z_{R_i}}{\sum_{i=1}^{n} Z_{H_i}}; \tag{7.15}$$

- formula (7.14) is the product of two predictions and is not optimal;
- two predictions obtained by different means are combined: $\tilde{R}$ is calculated with few laboratory samples (hundreds), while block kriging is obtained on the basis of a moving neighborhood selection among tens of thousands of measurements.

7.1.4 Tests

Let us remain in northern Chile in the underground Chuquicamata project and see if the usual practices have an important impact in the economic evaluation of the deposit and if cokriging really constitutes an improvement. A data set of 1,000 samples is available, in which Z_H, Z_T and Z_C are known together, distributed in three mineral zones. The domain covered by the samples is approximately $1 \times 4 \times 1$ km^3 (Figure 7.3).

The importance of additivity is assessed at two different scales: first, the scale of the production blocks and then the scale of the entire deposit. For the additivity at the scale of the blocks, the tests are carried out in mineral zone 409 (the best informed) in the central domain marked by a light gray square, measuring $200 \times 200 \times 50$ m^3 (Figure 7.3). This domain is divided into 500 blocks of $20 \times 20 \times 10$ m^3, where each block is finely discretized every 2 m. Z_H, r and Z_{TT} are cosimulated at each node of this grid. The size of the blocks is imposed by the production. The direct and cross-variograms (Figure 7.4) are fitted using an intrinsic correlation model established for

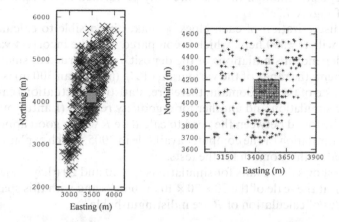

Figure 7.3 Area covered by laboratory samples. The gray square in the center of the area represents the domain of the tests based on conditional simulations.

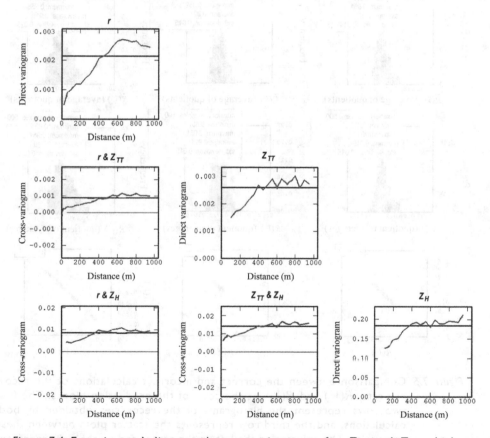

Figure 7.4 Experimental direct and cross-variograms of *r*, Z_{TT} and Z_H, which are further modeled for simulations.

the entire deposit, although the tests are only performed in the light gray domain in the center of Figure 7.3.

The fine discretization of each block V makes it possible to calculate the correct value (7.15), with $n = 500$ here, which is compared with the incorrect value (7.13). One hundred independent simulations of the deposit are constructed, since only one cannot be sufficient: the range of the variogram of Z_H (more than 300 m) is approximately equal to the size of the test domain; therefore, statistical fluctuations can be important between one simulation and another for ergodicity reasons (Lantuéjoul, 1991). A balance had to be found between the need to calculate R with a good approximation and the memory limitation of the computer available in 2005, which explains why the study was restricted to this domain for the tests.

Figure 7.5 shows the results for simulations n°1, 10 and 20 (they all lead to the same conclusions): at the scale of the $20 \times 20 \times 10$ m^3 blocks and with this spatial variability, the two modes of calculation of R are indistinguishable.

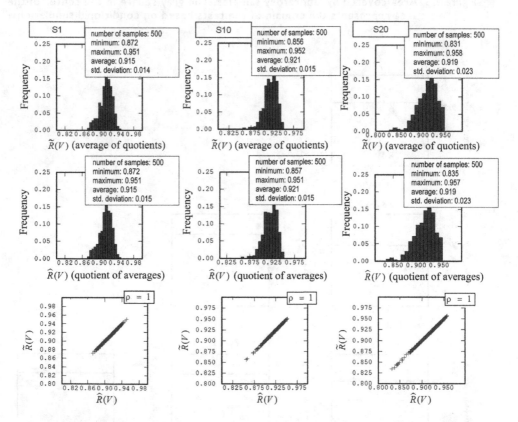

Figure 7.5 Comparison between the correct and incorrect calculations of the recovery rate, $\hat{R}(V)$ and $\tilde{R}(V)$, at the scale of the production blocks. The first two rows represent the histograms of the recoveries obtained by both calculations, and the third row represents the scatter plots between these recoveries. The columns on the left, in the center and on the right represent simulations n°1, 10 and 20, respectively.

Other experiments, at the same scale, where the predictor $\tilde{Z}_R(V_i)$ (formula (7.14)) is compared with a cokriging of Z_R with Z_H and r as covariates, show no more difference for reasons to be explained.

Before, it is worth asking: what happens at the scale of the entire deposit, when all the samples are used to evaluate the global metallurgical recovery? Table 7.1 shows the results for the three mineral zones.

At this scale, the deviations prove to be important. By systematically underestimating the recovery, the incorrect calculation can lead to a sizing of the smelter and of the refinery (here, adjacent to the mining site) that will not be adequate for the subsequent demand. Producing more concentrate than expected is always good news, but it is necessary that the smelter, which produces copper anodes (called blister copper) by a pyrometallurgical processing of the flotation concentrate, and the refinery, which produces copper cathodes by electrorefining the blister copper, are not prejudiced in terms of their capacity or performance.

In summary: at the scale of the deposit, additivity intervenes while, at the scale of a small block, it does not. Why?

A first reason: for the average of ratios...

$$\tilde{R} = \frac{1}{n}\sum_{i=1}^{n} R_i = \frac{Z_{R_1}}{n Z_{H_1}} + \ldots + \frac{Z_{R_n}}{n Z_{H_n}}$$

(7.16)

to differ little from a ratio of averages...

$$\hat{R} = \frac{\sum_{i=1}^{n} Z_{R_i}}{\sum_{i=1}^{n} Z_{H_i}} = \frac{Z_{R_1}}{\sum_{i=1}^{n} Z_{H_i}} + \cdots + \frac{Z_{R_n}}{\sum_{i=1}^{n} Z_{H_i}},$$

(7.17)

it is enough to have, for most of the points i of block V:

$$n Z_{H_i} \approx \sum_{i=1}^{n} Z_{H_i}.$$

(7.18)

Table 7.1 Global recovery at the scale of the entire deposit

Zone	Number of data	Variance of Z_H	Average of Z_H	Coeff. of variation (%)	$\bar{R}(V)$ Average of recoveries (incorrect) (%)	$\hat{R}(V)$ Quotient of averages (correct) (%)
409	671	0.18	0.91	47	88.8	89.3
207	394	0.32	1.16	49	88.8	89.8
206	47	1.21	2.22	49	86.3	86.9

The incorrect calculation systematically underestimates the recovery. For a deposit of this size, the subevaluation of the recovered metal represented 1.5 billion US dollars of 2005.

In other words, it is enough that each grade Z_{Hi} differs little from the average grade in the block, i.e., that the 'dispersion variance' $\sigma^2(\bullet|V)$ of the point-support grade within the block is small. This dispersion variance, well known to geostatisticians (Matheron, 1965a, 1971), is easily calculated if the variogram γ of the feed copper grade Z_H is known:

$$\sigma^2(\bullet|V) = \frac{1}{V^2} \int_V \int_V \gamma(x - x')\,\mathrm{d}x\,\mathrm{d}x' = \bar{\gamma}(V,V)$$

(7.19)

Let us examine the variogram of Z_H (Figure 7.6).

This variogram has been calculated on the entire deposit, but it also characterizes the domain of the tests, insofar as stationarity is assumed. As the distances for calculating $\bar{\gamma}(V,V)$ do not exceed 30 m, $\sigma^2(\bullet|V)$ is of the order of 0.1 and is close to the value of the nugget effect. Even if the local distribution of the grade varies from one block to another, for each block, the variation around the average in the block is small, and the coefficient of variation at the scale of the blocks is of the order of 35%.

If the size of block V now increases to that of entire deposit D, the dispersion variance of a point in a block becomes the variance of the sample in the deposit and increases to 0.18, the variogram sill. The coefficient of variation then increases to 47%, and the differences between the additive and nonadditive calculations become noticeable. As a consequence, things change in a relatively small range [0.35, 0.47] of the coefficient of variation.

A second cause relates to the variance of Z_R conditionally to Z_H. If, for each point i of block V, the ratio Z_{R_i}/Z_{H_i} varies little (irrespective of the variations of Z_{H_i}), the average of these ratios changes little and differs little from the correct calculation. Let us examine the cross-diagram between Z_H and Z_R for all the samples of the mineral zone 409 of the deposit (Figure 7.7).

The high correlation, equal to 0.995, indicates that Z_R follows Z_H linearly and that 99% of the variance of Z_R is explained by this linear dependence. The curve in Figure 7.7 shows that the conditional expectation of Z_R knowing Z_H is linear. The most

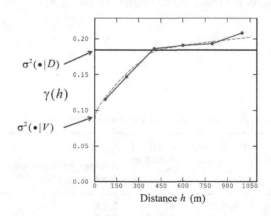

Figure 7.6 Experimental variogram of the feed copper grade Z_H. The bottom arrow indicates the variance of the grade of a sample in a block (nugget effect value), whereas the top arrow indicates the variance of this same grade in the deposit (varogram sill).

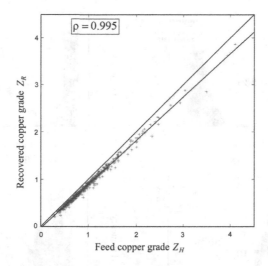

Figure 7.7 Cross-diagram between Z_H and Z_R for the samples of mineral zone 409.

important thing here is that the conditional variance of Z_R knowing Z_H is small whatever Z_H, although there is a slight tendency to increase with Z_H. Therefore, at each point i of V, in the Z_{R_i}/Z_{H_i} ratio, each value Z_{H_i} (which varies little in V) is associated with a value Z_{R_i} that varies little around Z_{H_i} multiplied by a factor 0.92, the slope of the regression of Z_R on Z_H (the intercept of this regression is almost zero). Five hundred stable ratios Z_{R_i}/Z_{H_i} are produced for the block, whose average $\tilde{R}(V)$ differs little from $\hat{R}(V)$, the ratio of the average values of Z_R and Z_H in V.

7.1.5 Conclusions

The fact that, at the block scale, the incorrect calculation has little influence cannot justify its use. It would suffice that the range of the variogram of the feed copper grades be smaller or that the variance increases, for the calculation errors to become important. The company found this demonstration convincing enough to definitively ban the incorrect calculation and to encourage the prediction of Z_R through a cokriging of Z_R, Z_H and r, as a precautionary principle.

This new approach is satisfactory for the economic geologist because it directly provides the desired variable Z_R, but not for the metallurgist, who wishes to evaluate the performance of the process by examining R. Now, regardless of how they are calculated, the maps of R show little variation (the numerator and the denominator are strongly correlated). In general, a quotient is rarely informative, since it does not give access to the causes that explain its variations. A spatial mapping aims at showing a regionalization implicitly linked with the geology. If two different spatialized causes lead to the same quotient, the link with the regionalization is lost. This was true in geotechnics, and it is true here. Figure 7.8 shows that the coefficient of variation of R is of the order of 5%, for initial quantities whose coefficient of variation is ten times greater or more.

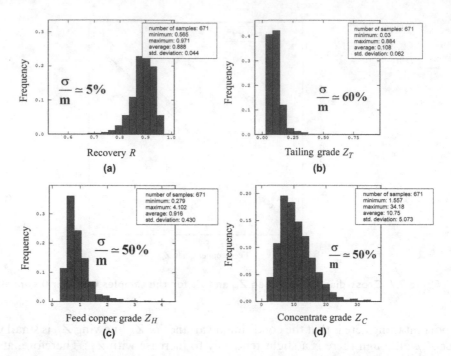

Figure 7.8 Characteristics of geometallurgical quantities. Histograms in the mineral zone 409. (a) Recovery R, (b) tailing grade Z_T, (c) feed grade Z_H, (d) concentrate grade Z_C.

Let us examine how R depends on Z_H, Z_T and Z_C.

In the ratio $(Z_H - Z_T)/(Z_C - Z_T)$ of formula (7.12) that delivers R, Z_H/Z_C is factored to further simplify R and to obtain the following first-order expansion:

$$
\begin{aligned}
\frac{Z_H - Z_T}{Z_C - Z_T} &= \frac{Z_H}{Z_C} \frac{1 - \dfrac{Z_T}{Z_H}}{1 - \dfrac{Z_T}{Z_C}} \\
&\approx \frac{Z_H}{Z_C} \left(1 - \frac{Z_T}{Z_H} \right) \left(1 + \frac{Z_T}{Z_C} \right) \\
&= \frac{Z_H}{Z_C} \left(1 - \frac{Z_T}{Z_H} + \frac{Z_T}{Z_C} - \frac{Z_T}{Z_H} \frac{Z_T}{Z_C} \right).
\end{aligned}
\tag{7.20}
$$

In Eq. (7.20), the approximation between the first and second lines is good since Z_T/Z_C is of the order of 1% on average and rarely exceeds 10%. In addition, even if the Z_T/Z_H ratio is not small, being of the order of 10% on average and sometimes beyond 50%, its product with Z_T/Z_C is insignificant, which leads to the following approximation for R, valid for this deposit and its variability:

$$
R \approx 1 - \frac{Z_T}{Z_H} + \frac{Z_T}{Z_C}.
\tag{7.21}
$$

There are several lessons to learn from this formula. For example, it is more interesting to decrease Z_T by 0.01, than to increase Z_C by 0.1 or 10%. If it is necessary to favor one over the other for technical reasons, the efforts should focus on the reduction of Z_T, as they will be more beneficial for the metallurgical recovery.

Another observation: Since Z_T/Z_H plays an essential role in the recovery, its variations must be studied correlatively with R, as well as the relative variations of Z_T and Z_H. Let us examine Figure 7.9.

Z_T and Z_H are linearly related with a slope that depends on R (Figure 7.9b). Therefore, if, on average, Z_T increases with Z_H, the rate of increase is stronger for low recoveries than for high recoveries. This is due to the conditional distribution of Z_H knowing Z_T/Z_H (Figure 7.9c). When Z_T/Z_H decreases, the dispersion of Z_H increases. Since R and Z_T/Z_H are negatively correlated (Figure 7.9a), high recoveries are associated with a wide range of grades Z_H, whereas low recoveries concern only low grades. This explains the change of slope in the cross-diagram 7.9b.

The aforementioned results apply to the first term Z_T/Z_H of formula (7.21). One should proceed identically for Z_T/Z_C and then study the relationship between Z_T/Z_H and Z_T/Z_C, with the idea of obtaining statistical simplifications that lead to direct conditional simulation methods for the metallurgical recovery. Geostatistics has pointed out the useful and influential variables; the geometallurgist must analyze them.

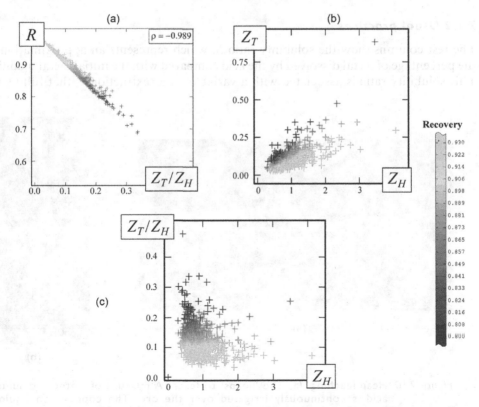

Figure 7.9 Example of analysis of geometallurgical quantities. Cross-diagrams (a) between Z_T/Z_H and R, (b) between Z_H and Z_T, (c) between Z_H and Z_T/Z_H.

7.2 Oxidized copper ore

7.2.1 Recovery by heap leaching

After being crushed into particles of about one centimeter in diameter and prepared in agglomerates (porous and mechanically stable material), the ore is deposited on a sealed floor to form a heap of the order of a dozen meters high and several hundreds of meters long and wide (Figure 7.10a) (Bartlett, 1997). On the ore and, sometimes, under protective covers that prevent evaporation, a mixture of water and sulfuric acid is spread, which dissolves the soluble minerals (copper oxides) and releases the metal in a continuous flow recovered at the base of the heap in channels or pipelines that bring the fluid to the solvent extraction and electrowinning plant. Heap leaching provides a metallurgical recovery rate R of the order of 70%, that is, on average ten points less than the recovery associated with flotation, although the leaching process is more economic as the acid is recirculated.

Laboratory experiments are carried out to evaluate the performance of the process. To this end, leaching columns of several meters high are used, containing ore in which the acid diffuses at a speed comparable with the one supposedly involved in the heaps (Figure 7.10b) (Bartlett, 1998). It takes several weeks to extract all the leachate by gravity. This explains the small number of available laboratory measurements, of the order of only 30 for the Gabriela Mistral deposit studied in the following.

7.2.2 Usual practices

The test columns show the solubility ratio S, which represents an approximation of the percentage of metal dissolved by the acid compared with the initial metal quantity. This solubility ratio is associated with a variable – the 'extraction' – that fulfills the

(a)

(b)

Figure 7.10 Heap leaching. (a) The 'heaps' to leach. A mixture of water and sulfuric acid is continuously irrigated over the ore. The copper-rich liquid is brought via pipelines to pools (top right) and then sent to the solvent extraction and electrowinning plant. (b) Laboratory leaching column.

role of the recovery R and is one of the products of the leaching column. For engineers, the goal is to derive R from S through a linear regression:

$$\hat{R}_{lab} = \hat{a}_{lab} + \hat{b}_{lab}\,S. \tag{7.22}$$

In addition, tens of thousands of samples are available in the deposit with information on the total in situ grade Z_H and the soluble grade Z_S, equal to the metal concentration contained in an acid solution applied to small samples. Since the grades Z_H and Z_S are related to the mass, supposedly constant, of samples that have the same size, these two additive quantities can be predicted by kriging on the support of a production block V of volume $10 \times 10 \times 10$ m^3. These predicted grades are then divided to obtain a solubility ratio at the support of block V:

$$\hat{S}(V) = \frac{Z_S^K(V)}{Z_H^K(V)}. \tag{7.23}$$

The metallurgical recovery at the block support is obtained by combining Eqs. (7.22) and (7.23):

$$\hat{R}(V) = \hat{a}_{lab} + \hat{b}_{lab}\,\hat{S}(V). \tag{7.24}$$

The metallurgist analyzes the performance of the process by studying $\hat{R}(V)$, whereas the geologist assesses the economic value of the block by calculating its recovered grade:

$$Z_{\hat{R}}(V) = \hat{R}(V)Z_H^K(V) = \hat{a}_{lab}Z_H^K(V) + \hat{b}_{lab}\,Z_S^K(V). \tag{7.25}$$

Apart from the usual questioning about the optimality and unbiasedness of this combination of predictors, an important difference with respect to the flotation of the sulfide copper ore stands out: the willingness to calculate a recovery for each block V, whereas in the other case, only one recovery rate is used for the entire deposit. Another observation that comes from the same pragmatism of the operators: in the previous expressions, and for the deposit that will be studied hereunder, the coefficient $\hat{a}_{lab}$ is not zero. Even if the measured solubility is equal to zero, the operator is considering a nonzero recovery of the metal. Seen from this angle, the expression (7.25) shows that it represents the percentage of the total grade that must be added to the soluble copper grade to obtain the desired quantity. Geometallurgists are aware that both the predicted solubility of Eq. (7.23) and the regression (7.22) modeling the recovery can be improved.

7.2.3 Distributions of the recovered grade and recovery rate

Still in the Atacama Desert, we move to the Gabriela Mistral deposit, located about 130 km southwest of Chuquicamata and in operation since 2008. Thirty laboratory measurements are available, providing S and R. The geographical coordinates of the samples are unknown; only their belonging to three zones (North, South and Periphery) is known.

In addition to these few data that provide R and are used to construct the linear regression (7.22), more than 15,000 drill-hole samples are available, in which Z_H and Z_S are measured, allowing the calculations (7.23), (7.24) and (7.25) (Figure 7.11).

Let us start with the laboratory and observe the first regression line, based on all the laboratory data, without distinguishing the geographical zone (Figure 7.12a).

For a zero solubility, the assumed recovery is $\hat{a}_{lab} = 0.29$, a quantity that considerably changes when the regression is carried out by geographical zone (Figure 7.12b–d, and table in the center of the figure). Although the number of samples per zone is too small to give credit to such regressions, it is clear that considering a single regression for all the zones is certainly not the best option. The conclusion that emerges once again is to predict $Z_H(V)$, $Z_S(V)$ and $Z_R(V)$ by cokriging or by joint simulation. With 15,000 point-support pairs (Z_H, Z_S) and only 32 triplets (Z_H, Z_S, Z_R) on a second larger support, one is in a highly heterotopic context where multivariate analyses should be favorable. Unfortunately, because the laboratory samples had unknown coordinates, these predictions could not be made.

Derived from the data in Figure 7.11, the coefficients of variation associated with Z_H, Z_S and S are 59%, 63% and 32%, respectively (Figure 7.13a–c), significantly higher than those of Z_H and R found in flotation (Figure 7.8a and c). This presages more important spatial variations for the recovery (extraction) R.

The linear regression of Z_S upon Z_H (Figure 7.13d) shows a significant correlation, greater than 0.7, and a significant dispersion around this regression. This reinforces the conclusions about cokriging or cosimulating (Z_H, Z_S, Z_R): the pair of coregionalized variables (Z_H, Z_S) is insufficient, and the third variable is essential to explain all the variability of Z_H.

By using the different regressions per zone to calculate Z_R according to formula (7.25) in which the volume V is not a block, but the volume associated with a sample, the distributions of Figure 7.14a are obtained, which can be compared with the one

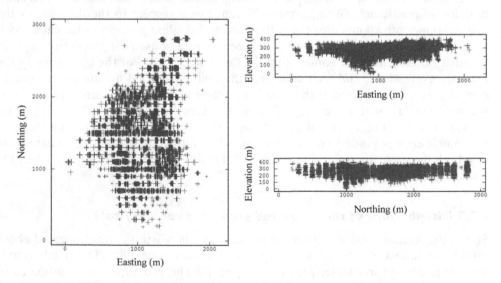

Figure 7.11 Gabriela Mistral deposit: 15,000 drill-hole samples in which the total copper grade Z_H and the soluble copper grade Z_S are known.

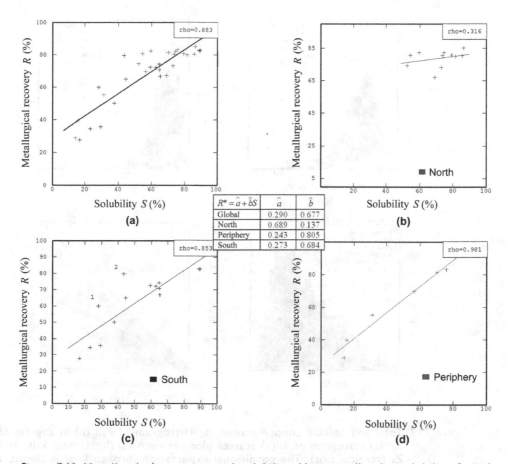

Figure 7.12 Metallurgical recovery and solubility. Horizontally, the solubility S, and vertically, the 'extraction' playing the role of the metallurgical recovery R. (a) All the samples are used. (b), (c), (d) Samples by zone, respectively, north, south and periphery. In the center, a table with the different regression coefficients, as per formula (7.22).

associated with a global regression. The calculation for the south zone is not possible, because there are not enough samples in this zone. The distributions of $Z_{\hat{R}}$ are very close in all the cases. The reason is as follows: the distributions of Z_H and Z_S are close (Figure 7.13a and b), and the correlation between these two variables is high (Figure 7.13d). Since the sum $\hat{a}_{lab} + \hat{b}_{lab}$ is always close to 1, a small contribution of one of the two variables Z_H and Z_S in formula (7.25) is compensated with a larger contribution of the other variable, with a result that ultimately depends little on the regressions.

Now, when examining the different distributions of the recovery R, calculated using formula (7.22) and the true value of S:

$$S = \frac{Z_S}{Z_H}$$

$$(7.26)$$

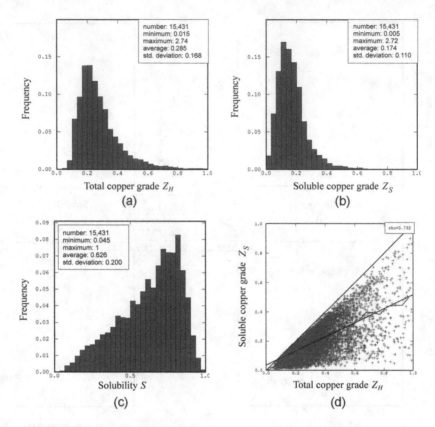

Figure 7.13 Total and soluble copper grades. (a) Histogram of Z_H; (b) histogram of Z_S; (c) histogram of S; (d) scatter plot between Z_H (horizontal axis) and Z_S (vertical axis). The conditional expectation (broken line) is almost a straight line.

important differences appear between these distributions, due to the strong dispersion of S around its regression (Figure 7.14b).

For the distribution associated with the north zone, R varies little around its average because in formula (7.24) that gives R as a function of S, the coefficient $\hat{b}_{lab}$ is about 0.14, for a constant $\hat{a}_{lab}$ of 0.7. The fluctuations of S around its regression are very little reflected in R for this zone, unlike the periphery zone where the opposite occurs ($\hat{a}_{lab}$ equal to 0.25 and $\hat{b}_{lab}$ greater than 0.8). Since most of the available samples belong to this peripheral zone, the global distribution is close to the distribution associated with this zone.

As a conclusion of this study, the small number of laboratory data did not allow obtaining reliable linear regressions, and the absence of coordinates made it impossible to cokrige Z_H, Z_S and Z_R. However, sensitivity tests show that, with the introduction of the solubility as a new actor, the situation is reversed with respect to flotation. While, in flotation, the metallurgical recovery rate R is relatively constant throughout the deposit, this coefficient here becomes extremely variable from one place to another.

It seems clear that the laboratory measurements should be repeated, with the idea of cokriging Z_H, Z_S and Z_R to optimally predict the recovered grade Z_R. This cokriging

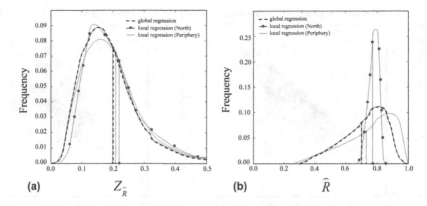

(a) $Z_{\widehat{R}}$ (b) $\widehat{R}$

Figure 7.14 Global and local regressions. (a) Distributions of $Z_{\widehat{R}}$ when regressions are used in each zone, compared with a global regression. (b) Distribution of the recovery rate $\widehat{R}$ per zone.

would jointly use the laboratory data (in which the metallurgical recovery is measured) with the numerous drill-hole data (in which the recovery is not available but the solubility is), which takes into account the different supports between the laboratory measurements and the drill-hole measurements.

7.2.4 Simulating the solubility on a block support

Formula (7.23) calculates the solubility of a block V by the ratio of the kriged soluble and total copper grades. This ratio is biased because it divides two predictions, unless the latter are made with exactly the same parameters (variogram model and kriging neighborhood) (Séguret et al., 2014, and Chapter 6, Section 6.2.1), which is seldom the case. In addition, the prediction is smooth compared with true solubility. On the other hand, formula (7.26) uses the grades measured in the drill-hole samples, thus avoiding the smoothing effect, but this formula is only valid for the quasi-point support associated with a sample and not for a voluminous block.

Simulating the block-support solubility would offer an alternative to formulae (7.23) and (7.26). However, such a simulation is complex due to the fact that the solubility is not additive: therefore, it is not enough to simulate the solubility at a point support and average it in a block, because this calculation would also be incorrect.

An easy-to-implement approach is presented hereunder to avoid this inconvenience (Emery et al., 2004), based on the joint simulation of the total copper grade (Z_H), the soluble copper grade (Z_S) and the solubility ratio (S). Since these three variables are linked by the relation $S = Z_S/Z_H$, it is necessary to simulate only two of them in order to derive the third one. The choice of these two variables is guided by the model used for the simulation, in this case, the multi-Gaussian model. First, the Z_H, Z_S and S data are transformed into Gaussian values, and the scatter plots between transformed variables are examined (Figure 7.15). Only the Gaussian transforms of Z_H and S draw an elliptical (almost circular) scatter plot that is compatible with a multi-Gaussian hypothesis. Therefore, this pair of variables will be chosen to perform the simulation. Although

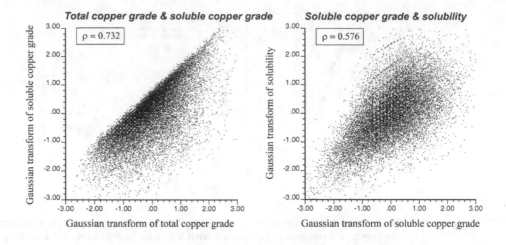

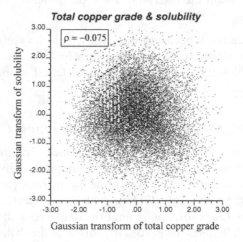

Figure 7.15 Scatter plots between the Gaussian transforms of the total copper grade, soluble copper grade and solubility. The scatter plot between the Gaussian transforms of the total copper grade and the solubility (bottom) is almost circular, which is the reason why this pair of variables is chosen for multi-Gaussian simulation.

both variables are poorly correlated, with a correlation coefficient equal to −0.075, the simulation will reproduce the correlations observed between the Gaussian transforms of Z_S and Z_H and between those of Z_S and S (0.732 and 0.576, respectively), given that Z_S is the product of the two simulated variables Z_H and S.

The low correlation between the Gaussian transforms of the total copper grade, and the solubility is corroborated by the variogram analysis of these two variables: the cross-variograms exhibit very low values compared with the direct variograms (Figure 7.16). In other words, the total copper grade has little influence on the solubility and, consequently, on the metallurgical recovery (eq. 7.22). The direct and cross-variograms are fitted with a linear model of coregionalization that includes a nugget effect, spherical and exponential

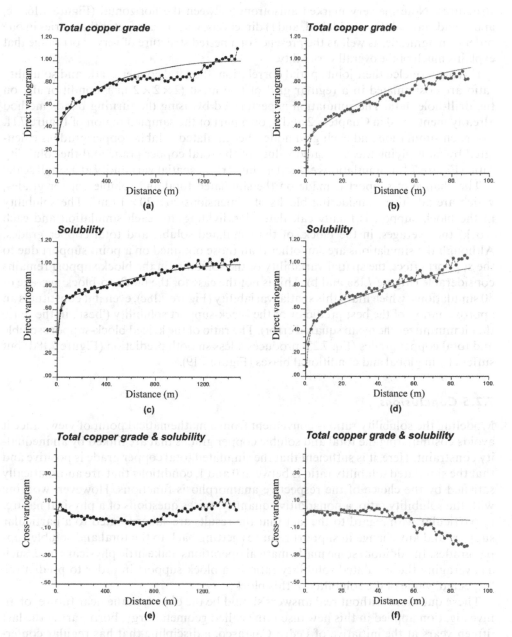

Figure 7.16 Direct and cross-variograms of the Gaussian transforms of the total copper grade and solubility ratio, along the horizontal (a, c, e) and vertical (b, d, f) directions. Dotted lines and dots: experimental variograms. Continuous lines: modeled variograms.

structures. Note the very marked anisotropy between the horizontal (Figure 7.16a, c, and e) and vertical (Figure 7.16b, d, and f) directions, with a scale factor that varies in one order of magnitude, as well as the presence of a nested structure of very short range that explains much of the overall variability.

Having modeled their joint spatial correlation, the total copper grade and solubility ratio are cosimulated in a regular grid of fine mesh ($2 \times 2 \times 2$ m^3), conditionally on the drill-hole data. The simulation is performed by using the turning bands method already mentioned in Chapters 2 and 4, on a part of the sampled region of Figure 7.11. For each simulation and each grid node, the simulated soluble copper grade is calculated by multiplying the simulated values of the total copper grade and the solubility ratio (Figure 7.17). Simulations exhibit an important spatial variability at a small scale.

The change of support is made on the simulated total and soluble copper grades, which are additive, considering blocks of dimensions $20 \times 20 \times 10$ m^3. The solubility in the block support is finally calculated by dividing, for each simulation and each block, the averages, in the block, of the simulated soluble and total copper grades. Although the simulations are smoother than those obtained on a point support due to the support effect, the spatial variability of the solubility in the block support remains considerable (Figure 7.18a and b). This is not the case for the block-to-block average of 50 simulations, which hides this spatial variability (Figure 7.18c), even if it constitutes an approximation of the best predictor of the block-support solubility ('best', in the sense that it minimizes the mean squared error). The ratio of the kriged block-support soluble and total copper grades (Eq. 7.23) produces a less smooth prediction (Figure 7.18d) but suffers from global and conditional biases (Figure 7.19).

7.2.5 Conclusions

Modeling the solubility ratio is convenient from a mathematical point of view, since it avoids cosimulating the total and soluble copper grades that are linked by an inequality constraint. Here, it is sufficient that the simulated total copper grade is positive and that the simulated solubility ratio is between 0 and 1, conditions that are automatically satisfied by the choice of the respective anamorphosis functions. However, working with the solubility ratio, a nonadditive quantity, raises questions of a physical nature, in particular with regard to the fact that the results are subordinated to a particular support and any change in support requires getting back to the total and soluble copper grades. In addition, some mathematical operations make little physical sense, such as averaging the simulated solubility ratio in a block support in order to predict the 'expected' value of the solubility of this block.

These questions without real answers should be the object, in the near future, of an investigation applied in this new discipline called geometallurgy, born during the last fifteen years at the initiative of Pedro Carrasco, a discipline that has regular conferences in Latin America (2012, 2014 and 2017-2020 in Santiago de Chile, 2016 and 2018 in Lima, Peru), Australia (2011 and 2013 in Brisbane, 2016 in Perth) and South Africa (2018 in Cape Town) – keyword 'geomet'. Note that the first congress in Brisbane in 2011 was dedicated to Pedro Carrasco, the initiator of the fruitful collaborations of which this work is, in part, the summary (Beniscelli, 2011).

Geometallurgy is now taught in universities on five continents. Some recently addressed research topics are the influence of mineralogy on the economic evaluation of

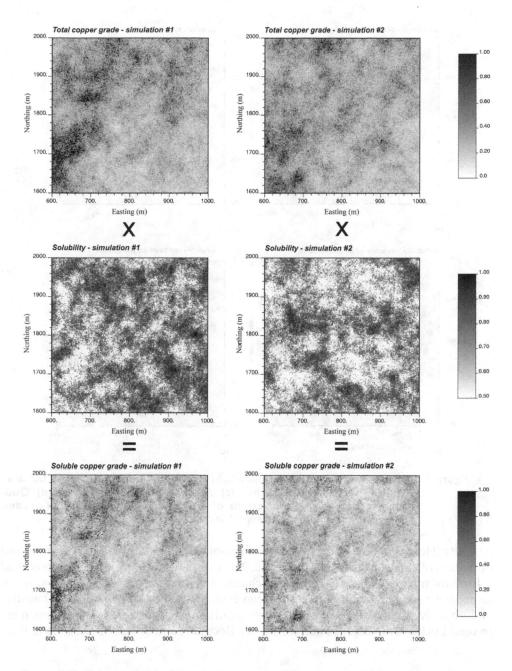

Figure 7.17 Two conditional simulations on a point support of the total copper grade (a), solubility ratio (b) and soluble copper grade (c), derived by multiplying the total copper grade by the solubility ratio. Grades are measured in percentages; the solubility ratio is a fraction between 0 and 1. Representations of a horizontal section of a sector of the deposit.

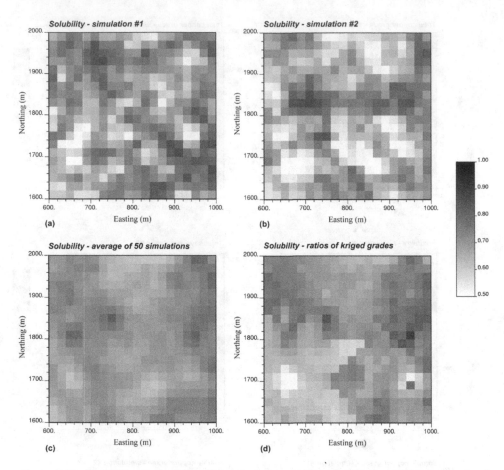

Figure 7.18 (a, b) Two conditional simulations of the solubility ratio on a block of dimensions $20 \times 20 \times 10$ m^3. (c) Average of 50 simulations. (d) Quotient of the kriging predictions of the soluble and total copper grades. Representations of a horizontal section.

deposits (Hoal et al., 2013), the influence of geological uncertainty in the planning and operation of concentrators (Navarra et al., 2017), or the incorporation of hyperspectral information from drill-hole samples to characterize their mineralogy and texture or to predict rock grindability indices (Linton et al., 2018). For an apparently exhaustive and objective historical review of the geometallurgical approach, it is recommended to read Lotter et al. (2011) as well as van den Boogaart and Tolosana-Delgado (2018).

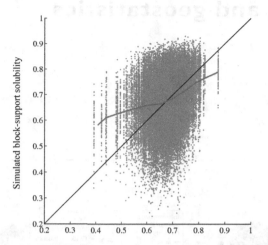

Block-support solubility predicted by quotient of kriged grades

(a)

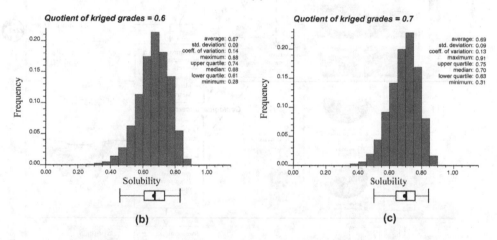

(b)

(c)

Figure 7.19 Conditional bias of the solubility predictor (7.23) based on the ratio of kriged block-support grades. (a) Scatter plot between the solubility predicted by Eq. (7.23) (horizontal axis) and the simulated solubility (vertical axis); the regression curve (gray broken line) differs from the diagonal (black straight line). (b) Distribution of simulated solubility when the prediction (7.23) is equal to 0.6; the average simulated solubility is then 0.67. (c) Same exercise with a prediction (7.23) equal to 0.7; the average simulated solubility is then 0.69, and the conditional bias is low for this specific value of the solubility predicted by Eq. (7.23).

Chapter 8

Sampling and geostatistics

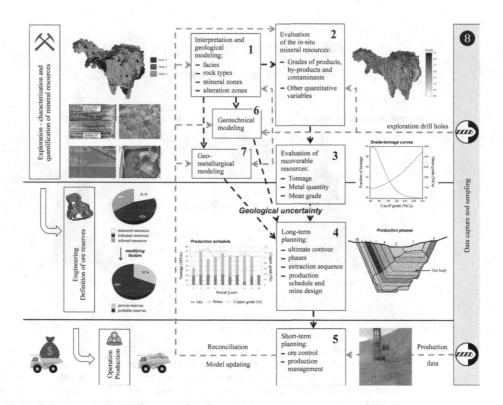

The term sampling, as studied by Pierre Gy from the 1950s (Gy, 1954, 1956), should be taken in a more general sense where the reality, at different scales, is known only through samples, the number and quality of which may vary. The first section of this chapter identifies the most common sources of sampling errors and presents some exploratory tools to check the quality of a set of sampling data. The next two sections focus on sampling biases related to the use of different sources of information and to a nonrepresentative (preferential) sampling of the region under study.

8.1 Information quality

The available exploration drill-hole data or production data can be subject to numerous errors that affect their coordinates or their values. These errors can occur in different sampling stages, in particular:

- registration of drill-hole headers;
- registration of drill-hole orientations;
- control of drill-hole deviations;
- collection or preparation of samples;
- identification, by the geologist, of the rock type, mineral type and degree of alteration;
- chemical analysis of grades;
- transcription and registration of the grade values.

One consequence is the occurrence of inconsistencies in the data sets provided to the geostatistician. For example, he/she could find data located above the topographic surface, high copper grades in areas that the geologist has logged as leached material, soluble copper grades greater than total copper grades, proportions of minerals that sum to more than 100% – sometimes a real time-consuming nightmare in a context where the study time is scarce. Before the geostatistical study, some precautions are required to minimize the sampling errors, including the following:

- registering and using the original geological logging of the drill-hole cores;
- unifying the criteria for the geological interpretation of the rock types, mineral types, alterations and structural domains;
- monitoring the performance of the drilling in relation to the drilled length, the control of deviations and the sample recovery;
- establishing quality assurance and quality control (QA/QC) systems with, in particular, the insertion of blanks (waste samples with zero or insignificant grades), standards (certified reference materials with known grades) and duplicates (two subsamples extracted from the same sample) in each batch of samples sent to the chemical analysis laboratory, in order to check the absence of contamination of the measurements, their accuracy (absence of systematic error or bias) and their precision (low error variance);
- establishing a policy that aims at preserving the data integrity.

Likewise, the geostatistician must carry out a thorough exploratory study of the data before moving on to the modeling, prediction or simulation stages (see Section A.1 in Appendix). In particular, this study may reveal some inconsistencies between the different sources of available information:

- interpreted geological model incompatible with the logging made on drill-hole samples;
- geological interpretation incompatible with the measured grades;

- spatial coordinates incompatible with the topographic survey;
- inconsistencies between grade measurements made through different types of sampling: diamond drill holes, reverse circulation drill holes, blast holes, among others.

Among the set of exploratory tools, three are particularly helpful for the detection of errors or anomalies: maps, (lagged) scatter plots and experimental variograms. We illustrate with the Codelco-Andina data presented in Chapter 3 (Figure 3.2). In addition to the total copper grades, the predominant rock type is recorded for each composite sample, grouped into three main facies: granodiorite (granitoid), highly mineralized tourmaline breccia and poorly mineralized contact breccia. The projection maps suggest several errors in the rock type record, with two drill holes and a few isolated samples labeled as contact breccia in the middle of the tourmaline breccia and granodiorite (Figure 8.1); these suspicious drill holes and samples may need to be relogged by geologists.

The copper grade measurements have been subject to a quality control program that includes the use of duplicates of the same drill-hole cores. The two grades measured on each pair of duplicates have been included in the database. Among a total of 72 pairs of duplicates, more than one half (37) have a difference of grades less than 0.1%; only two pairs of duplicates have measurements that differ by 1% or more (Figure 8.2), suggesting that all the samples belonging to the same batches should be reassayed. The semivariance of the errors (moment of inertia of the scatter plot) is equal to 0.036, a value very close to the modeled nugget effect of the variogram of the drill-hole grades (Figures 3.7a and 8.4b) and equal to one-tenth of the grade variance (Figure 3.3b). This order of magnitude of the measurement error dispersion comparatively to the

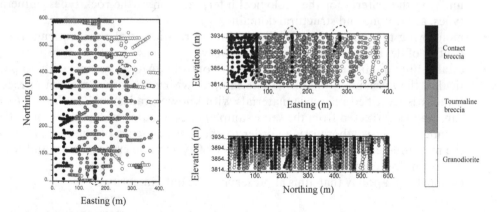

Figure 8.1 Data from the Sur-Sur mine of Codelco-Andina. Projections in horizontal and vertical sections of the samples in local coordinates. The scale indicates the rock type. Several data (enclosed in dotted ellipses) labeled as belonging to the contact breccia are isolated in the middle of the tourmaline breccia and granodiorite, suggesting errors in the rock type registration and the need for geological relogging.

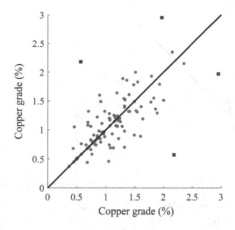

Figure 8.2 Data from the Sur-Sur mine of Codelco-Andina. Scatter plot between duplicate measurements of diamond drill holes. The squares indicate two pairs of measurements that differ by 1% or more. The semivariance of the errors between duplicate measurements (inertia of the scatter plot) is equal to only one-tenth of the copper grade variance.

copper grade dispersion is considered acceptable and corroborates the precision of the drill-hole measurements.

Finally, blast holes are drilled during production, with a support comparable with that of the drill holes (12 m in length) (Figure 8.3). Their average grade (1.15%) is much higher than that of the drill holes (0.90% if one refers to the declustered histogram in Figure 3.3b). However, this difference is not due to a bias in the blast-hole measurement

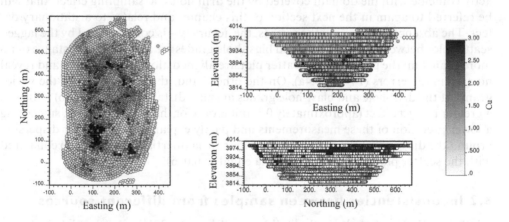

Figure 8.3 Data from the Sur-Sur mine of Codelco-Andina. Projections in horizontal and vertical sections of blast holes in local coordinates. The scale that indicates the copper grade is the same as that used in Figure 3.2.

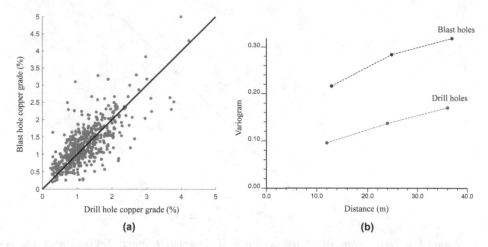

(a)   (b)

Figure 8.4 Data from the Sur-Sur mine of Codelco-Andina. (a) Lagged scatter plot
between drill-hole measurements and blast-hole measurements for a separation distance less than 4 m. The tendency of the points to align around
the diagonal indicates an absence of bias between the two types of measurement. The greater dispersion, compared with the scatter plot in Figure
8.2, is due to the lower precision of the blast-hole measurements. (b)
Experimental variograms calculated along the vertical direction for distances between 12 m and 36 m. The significantly greater nugget effect for
the blast holes corroborates the suspicion of its poorer precision, for two
variograms that could be derived from each other by a simple translation
parallel to the vertical axis. The supports of the two types of measurement are comparable (drilling length of 12 m).

but can be explained by the fact that the domain covered by the blast holes does not exactly coincide with the domain covered by the drill holes, a 'sampling effect' that will
be referred to again in the next section of this chapter and relates to a stationary de-
fect. The absence of bias – in other words, the accuracy – is corroborated by the lagged
scatter plot between the drill-hole and blast-hole grades for a separation distance of no
more than 4 m: the points of the scatter plot are aligned along the diagonal and reveal
no systematic errors (Figure 8.4a). On the other hand, the comparison of the variograms of the drill-hole and blast-hole grades at small distances (Figure 8.4b) indicates
a greater nugget effect (approximately 0.1 unit more) for the blast-hole data, suggesting
a lower precision of these measurements and partly explaining the greater dispersion
around the diagonal of the scatter plot in Figure 8.4a (inertia equal to 0.105) compared
with the scatter plot in Figure 8.2 (inertia equal to 0.036).

8.2 Inconsistencies between samples from different sources

In the study presented in the following (Séguret, 2011b), also related to the
Codelco-Andina mine in central Chile, laboratory data from geometallurgical samples (which measure the metallurgical recovery, among other variables) are compared
with exploration drill-hole data. The analyses show very informative inconsistencies
about procedures that may be common in the industry.

8.2.1 Problem setting

Based on the same original copper ore, the 'drill-hole' data and the 'laboratory' data (to use the expression of the professionals) should provide the same grades in the same places when the support is the same. Here, however, this is not the conclusion that is reached: the grades do not have the same range (histograms of Figure 8.5a and b) and the variograms are very dissimilar. In Figure 8.5c, the sill of the variogram of the laboratory data is less than one half of the sill of the variogram of the drill-hole data. It is almost a pure nugget effect for the laboratory data, while there is a spatial structure for the drill holes. To simplify the notations, the 'H' as head of the previous chapter is here replaced by 'T' as total.

What happened? Four guilty practices were identified by the 'geoinvestigator':

- spatial restriction;
- regularization;
- sampling density;
- grade selection.

8.2.2 Investigation details

Figure 8.6 shows that the domain covered by the laboratory data is smaller than the domain covered by the drill-hole data.

When the analysis of the drill-hole data is restricted to the domain covered by the laboratory data, the variance is reduced by 20% (Figure 8.7b). The difference is due to an isolated group of high values that do not belong to the same geological unit (facies), the grades being only approximately stationary at the kilometer scale. In the following, the analyses of the drill-hole data are limited to the domain covered by the laboratory data.

A second cause is the support of the measurements. The laboratory data have a 15 m support and the drill-hole data have a 10 m support, while the drilling diameters range from a few centimeters to tens of centimeters. Using the same deconvolution–convolution techniques as in Chapter 5 when comparing the blast-hole measurements with the drill-hole measurements, this support difference is estimated to cause a drop of less than 10% of the initial sill (Figure 8.8).

The third cause has more important consequences. The spatial domain covered by the laboratory data contains more than 3,000 drill-hole samples but only about 200 laboratory measurements. To assess the effect of the sampling density on the quality of the variogram, the initial set of drill-hole measurements is randomly divided into two subsets, trying to cover the domain in a homogeneous manner. The operation is performed three times to obtain 16 subsets of approximately 200 drill-hole samples. For each of these subsets, a variogram is calculated. Figure 8.9a and b shows that, with 800 samples, a variogram that characterizes the domain can still be calculated. The possible range of variograms becomes extremely wide with 400 samples (Figure 8.9c), and even frightening when only 200 samples are used (Figure 8.9d). In other words, the 200 laboratory samples are insufficient to reflect the spatial variability of the grades. Any of the 16 variograms in Figure 8.9d is admissible, with a range of possible nugget effects varying from simple to sixfold.

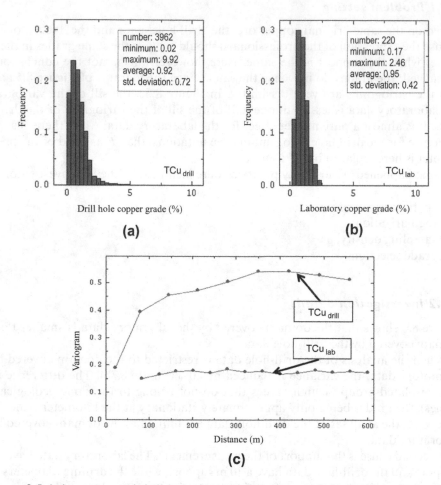

Figure 8.5 Laboratory and drill-hole data. (a) Histogram of the total copper grades TCu_drill of the drill-hole data. (b) Histogram of the total copper grades TCu_lab of the laboratory data. (c) Variograms of the drill-hole and laboratory grades.

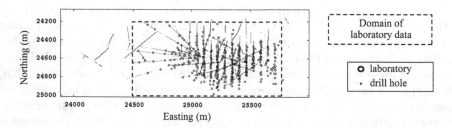

Figure 8.6 Codelco-Andina data. The domain covered by the drill holes is larger than that covered by laboratory data, in a context where the grades are not stationary at the scale of 1 km.

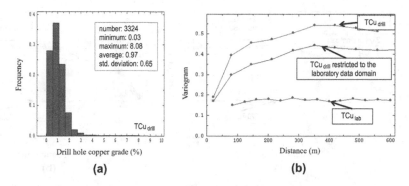

(a)   **(b)**

Figure 8.7 Effect of the domain covered by the measurements. (a) Distribution of drill-hole data in the domain of the laboratory data. (b) Variogram of the drill-hole data throughout the entire domain and in the laboratory data subdomain.

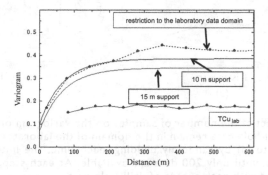

Figure 8.8 Effect of the change of support. The lower continuous variogram is the equivalent of the upper one when the support changes from 10 m to 15 m.

The last cause deals with the grade selection. The histogram in Figure 8.5b shows that the grades analyzed in the laboratory do not exceed 2.45%, while the drill-hole grades can exceed 9%. The same applies to the minimum, which, in the drill holes, can reach 0%, while it is not less than 0.17% in the laboratory. In fact, after a tense but polite interrogation, the laboratory operator confessed to the geoinvestigator that he discarded the extreme samples because he was looking for samples that were 'representative' of an average behavior, at the scale of a block, so that his metallurgical recovery rate could be used directly in the geologist's block model. Therefore, not all the variability of the total grades is sampled, nor the variability of the recovered grades that are related to the total grades. The predicted metallurgical recovery is probably very optimistic, which is why, after these calculations, the operators apply a 'safety factor' equal to 80% of the recovered grades calculated from these samples.

When the selection and trimming of drill-hole data are applied, a dramatic drop of the variance is observed (Figure 8.10a). In addition, a decrease in the amplitude of the set of possible variograms occurs in the case of a subsampling (Figure 8.10b).

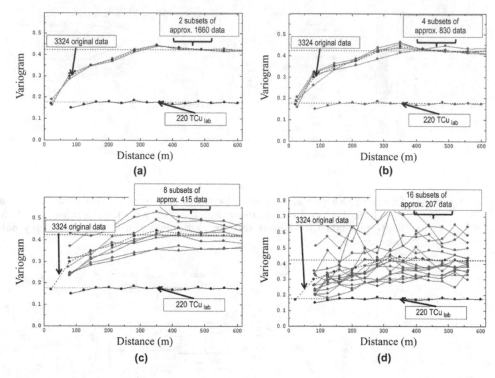

Figure 8.9 Effect of the number of samples on the variogram of drill-hole grades. The drill-hole data remain in the domain of the laboratory data. One goes from (a) to (d) by randomly dividing, each time, the initial number of data by two, until only 200 data are available. At each step, a variogram is calculated with each subset of drill-hole data.

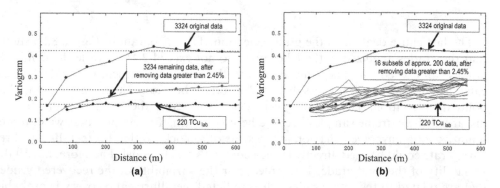

Figure 8.10 Effect of selection and trimming. (a) Intermediate curve: the variogram of the drill hole grades when data greater than 2.45% are omitted. Upper curve: the variogram of all the drill-hole grades in the laboratory domain. (b) The intermediate curves represent the different variograms obtained with approximately 200 drill-hole samples taken at random and without the grades greater than 2.45%.

8.2.3 Synthesis

Figure 8.11 summarizes the different causes involved, all of them related to sampling, the most significant one being, by far, a preferential grade selection, followed by the small number of data. Between 400 and 800 measurements are needed to cover the domain and to obtain a representative variogram of the grade variability.

8.2.4 Modeling the sampling effect

Since the regularization only represents a small percentage of the 'sampling effect', the deconvolution techniques used in Chapter 5 are not very useful in this case. It is preferred to use a direct multivariate approach to construct a model that allows the joint use of laboratory data and drill-hole data.

In order to evaluate the link between the total copper grades TCu_{drill} and TCu_{lab}, the process of migration is still necessary because no laboratory data exactly matches the location of a drill-hole data. Figure 8.12a shows the scatter plot obtained for data migrated less than 5 m. The variogram analysis leads to a linear model of coregionalization based on three independent nested structures: two structures, one for each variable, representing the measurement errors and a third, shared, structure giving an experimental correlation equal to 0.82. Established on the basis of migrated data, this model is then adapted to all the drill-hole data by introducing a scaling coefficient to account for the difference of variances between the TCu_{drill} variogram and the TCu_{lab} variogram of Figure 8.5c. The fitting of Figure 8.12b is obtained as the result.

It remains to link this model to the objective, which is to predict at each point of a fine grid the recovered copper grade RCu_{lab} measured in the laboratory. The correlation between RCu_{lab} and TCu_{lab} is almost perfect (Figure 8.12c), so that knowing one of these two variables is practically equivalent to knowing the other one. Accordingly, cokriging the three variables TCu_{lab}, RCu_{lab} and TCu_{drill} is not necessary, and a bivariate model (TCu_{drill}, RCu_{lab}) is enough. This is further justified by the fact that a linear regression of RCu_{lab} on TCu_{lab} delivers an unstructured residual whose variance is almost zero. This regression makes it possible to link RCu_{lab} with TCu_{lab}, therefore RCu_{lab} with TCu_{drill}, and leads to the fitting of Figure 8.13.

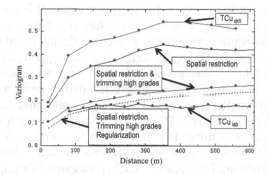

Figure 8.11 Synthesis of the different causes of the 'sampling effect'.

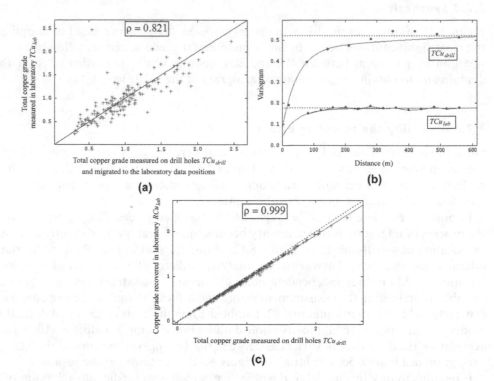

Figure 8.12 Total and recovered copper grades. (a) The scatter plot between TCu$_{drill}$ measurements (horizontal axis) migrated to the TCu$_{lab}$ measurements (vertical axis) located less than 5 m away. (b) Joint variogram fitting of TCu$_{drill}$ and TCu$_{lab}$. (c) Scatter plot between TCu$_{lab}$ (horizontal axis) and RCu$_{lab}$.

This model is used to predict the recovered grade RCu$_{lab}$ measured in the laboratory from the TCu$_{drill}$ and RCu$_{lab}$ data. To test it, a cross-validation is carried out on the laboratory data. The performance is assessed by the standard deviation of the errors between the true RCu$_{lab}$ values and their predictions. The histogram in Figure 8.14b shows that a cokriging based on the two types of measurements is significantly more efficient than a kriging based solely on the laboratory data (Figure 8.14c), since the error variance decreases by almost 50%. Although they are already good, these results can be further improved by predicting the recovered grade measured in the laboratory by a simple linear regression applied to the kriging of TCu$_{drill}$ (histogram of Figure 8.14a), regression based on the coefficients derived from the scatter plot of Figure 8.12c. The reason is twofold. First, in a context where the drill-hole measurements are dense, the cross-validation implies that, near each RCu$_{lab}$ measurement where a kriging of TCu$_{drill}$ is carried out, there is a TCu$_{drill}$ measurement, which receives a large kriging weight due to its proximity. This makes the test not much discriminatory since everything happens almost as if the predictor were equal to this nearby value. Second, cokriging RCu$_{lab}$ knowing RCu$_{lab}$ and TCu$_{drill}$ relegates TCu$_{drill}$ to the role of an auxiliary variable, less influential than the primary variable even if some auxiliary data

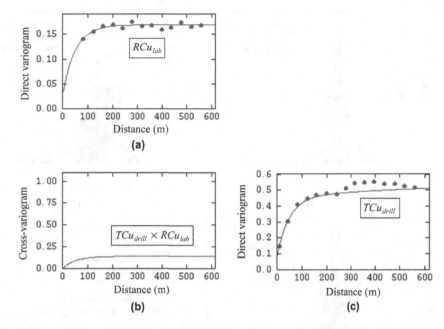

Figure 8.13 Final bivariate model. The points represent experimental variograms, and the continuous lines represent the fitted models. (a) RCu_{lab} variogram. (b) Cross-variogram between TCu_{drill} and RCu_{lab}: only the model is represented due to the total heterotopy of these measurements that prevents the calculation of the experimental cross-variogram. (c) TCu_{drill} variogram.

receive significant weights. A cokriging with related means (Emery, 2012a), in which both variables play the same role, would undoubtedly improve the results and make the performance of cokriging at least equal to that of the regression upon the kriging results. The proof of this statement will be made in the next section.

In short, the density of drill-hole measurements is so large compared with the laboratory data that the incorporation of the laboratory information adds almost nothing to the cokriging system, as was the case presented in Chapter 5 where the cokriging of blast-hole (many) and drill-hole (fewer) data was barely more efficient than a kriging based solely on the blast-hole data. Here again, the abundance of information prevails over the geostatistical model, whose purpose is to compensate for the lack of measurements.

8.3 Preferential sampling

Preferential sampling is another problem commonly found in the evaluation of mineral deposits. Two practices are particularly frequent.

The first is to intensify the sampling effort in the areas of the deposit with the highest grades. We had the opportunity to learn about this practice in Chapters 1 and 3 for the Río Blanco-Los Bronces deposit. In such cases, the experimental distribution of the sampled grades is likely to give a biased estimate of the underlying true distribution. The geostatistician generally applies compensation techniques, by assigning

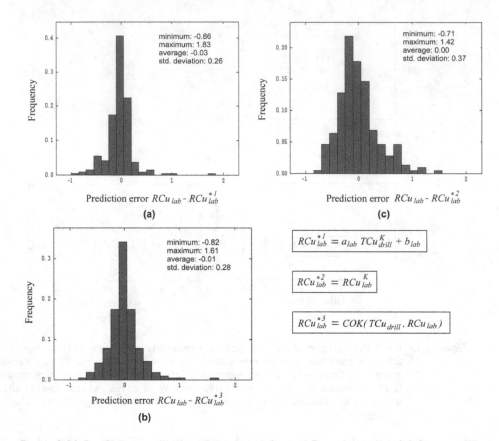

Figure 8.14 Prediction errors and cross-validation. Distributions of RCu_{lab} prediction errors resulting from cross-validation. The quality criterion of the prediction is the standard deviation of the errors. (a) The predictor is a linear regression on the kriging of TCu_{drill}, with coefficients established in the laboratory. (b) The predictor is a cokriging of RCu_{lab} and TCu_{drill}. (c) The predictor is a kriging of RCu_{lab}.

a greater weight to the data located in the subsampled areas and a lower one to the data of oversampled areas. The weights can be calculated by using algorithms based on purely geometric criteria (see, for example, Figure 3.3 in Chapter 3) or including the spatial continuity of the grades, for example, kriging weights (Chilès and Delfiner, 2012). Attention should be drawn to the fact that not only the experimental distributions are biased, but also the experimental covariances and variograms, the calculation of which should also be based on a weighting of the data (Rivoirard, 2001; Emery and Ortiz, 2007), a practice that unfortunately is still not widespread in 2020. In particular, when low-grade areas are undersampled, biases in the experimental variogram can result in a change of shape, an increase of the sill or the emergence of a hole effect with respect to the true underlying variogram (Emery, 2012b).

The second practice is to nonpreferentially sample the field when measuring the total grade and to measure another variable of interest (for instance, the soluble grade) only if the total grade is greater than a given threshold. This time, the bias affects the soluble grade distribution, its direct variogram and its cross-variogram with the total grade. Again, weighting techniques can be implemented to eliminate, or at least to mitigate, the bias, which can also be corrected by an ad hoc modeling of the bivariate distribution of total and soluble grades. This last approach is illustrated in the following in the case of the Gabriela Mistral deposit presented in the previous chapter.

From the data set in Figure 7.11 (more than 15,000 drill-hole samples with information on both the total and soluble copper grades, without preferential sampling), a new data set giving a biased image of the soluble copper grade distribution is created by eliminating the information on the soluble copper grade for 85% of the data with a total copper grade less than 0.3%, which mimics preferential sampling designs commonly found in mineral resources evaluation (Figures 8.15 and 8.16a). Based on this

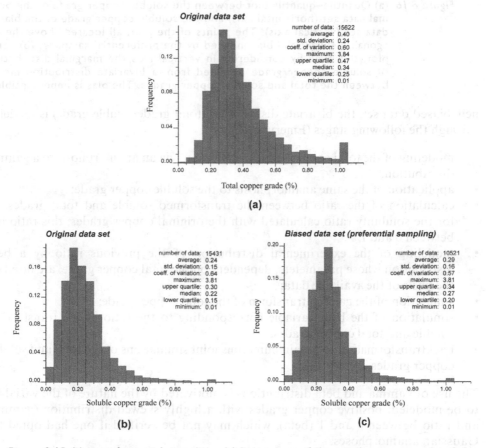

Figure 8.15 Nonpreferential sampling: (a) histogram of the total copper grade of the original data set (15,622 data), (b) histogram of the soluble copper grade of the original data set (15,431 data). Preferential sampling: (c) soluble copper grade histogram of the subsampled data set (10,521 data), which provides biased statistics compared to the histogram (b).

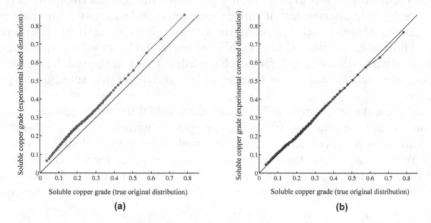

Figure 8.16 (a) Quantile–quantile plot between the soluble copper grade of the original data set (horizontal axis) and the soluble copper grade of the biased data set (vertical axis). The points of the plot, all located above the diagonal, highlight the bias induced by the preferential sampling. (b) Same plot, this time by considering in vertical axis the marginal distribution of soluble copper grade obtained from a bivariate distribution model between the total and soluble copper grades. The bias is imperceptible.

new biased data set, the bivariate distribution (total grade/soluble grade) is modeled through the following stages (Emery, 2012b):

- modeling of the total copper grade distribution, via an anamorphosis to a gamma distribution;
- application of the same anamorphosis to the soluble copper grade;
- calculation of the ratio between the transformed soluble and total grades; as for the solubility ratio calculated with the original copper grades, this ratio lies between 0 and 1;
- modeling of the experimental distribution of the previous ratio by a beta distribution whose parameters, dependent on the total copper grade, are fitted on the basis of the available data;
- simulation of the gamma transform of the total copper grade;
- simulation of the beta variable corresponding to the ratio of the transformed soluble and total copper grades;
- backtransformation to obtain numerous joint simulations of the total and soluble copper grades.

The use of gamma and beta distributions is motivated by the nature of the variables to be modeled: positive copper grades with a highly skewed distribution (gamma) and ratio between 0 and 1 (beta), which may not be verified if one had opted for Gaussian anamorphoses.

The distributions of the simulated total and soluble copper grades almost perfectly coincide with those observed in the original data set: in particular, the bias is no longer noticeable for the soluble copper grade (Figure 8.16b). The underlying hypothesis of the model that explains this success is that, for a given total copper grade, the experimental distribution of the soluble copper grade is unbiased. In other words, the preferential nature of the sampling distorts the prior distribution of the soluble copper grades, but not their distribution conditional on a known total copper grade.

Having corrected the marginal distribution of the soluble copper grade, one still has a long way to go through the geostatistical modeling and spatial prediction. In particular, biases also occur in the variogram analysis stage, as seen in Figure 8.17 when comparing the experimental variograms of the original and biased data sets. The latter set leads to a variogram with a greater nugget effect and a greater sill, i.e., it exhibits too much spatial variability at both short and large scales. This is explained because the low soluble copper grade measurements (which are removed in the biased data set) are located in areas with less variability, a phenomenon known as a 'proportional effect' (Chilès and Delfiner, 2012).

To make things worse, it is not enough to be able to infer without bias the distribution and the variogram of the soluble copper grade data; their interpolation by kriging can produce biased predictions even if the kriging parameters (mean and covariance or variogram) are estimated correctly. Indeed, one implicit postulate of kriging is that the choice of the sample positions does not depend on their values, a postulate that does not hold in the case of a preferential sampling.

When the total grade is not sampled preferentially, it is possible to use cokriging in order to remove or, at least, to attenuate the biases in the prediction of the soluble grade, provided that attention is given to the modeling of the mean grades. Indeed, traditional ordinary cokriging, which considers the means as unknown and unrelated, assigns to the data of the covariate (total grade) weights that add to zero (Wackernagel, 2003). As a result, the influence of the covariate on the prediction of the variable of interest (the soluble grade) is small and does not compensate for the sampling bias. If, on the contrary, a cokriging with related means is used, assuming that the mean soluble grade is a fraction, to be determined, of the mean total grade, these two means being unknown and possibly taking different values from one moving neighborhood to another, then the influence of the covariate is much more pronounced: the sampling

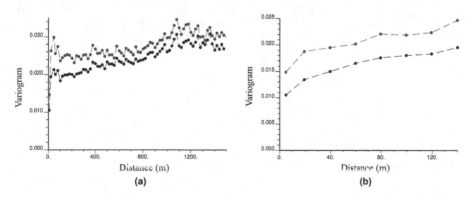

Figure 8.17 Experimental variograms of the soluble copper grades, calculated (a) in the horizontal plane and (b) along the vertical direction. The variograms with the lowest values correspond to the original data set (15,431 data). The variograms with the highest values correspond to the biased data set (10,521 data).

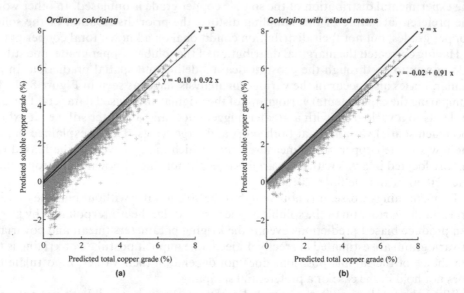

Figure 8.18 Scatter plots and regression lines between cokriged total and soluble block-support copper grades. The relationship between the grades improves considerably through the use of cokriging with related means (b) in comparison with the traditional ordinary cokriging (a). In particular, the intercept of the regression line is very close to its ideal value (zero), the negative predicted grades are scarcer and the inequality constraint between the total and soluble grades is almost always reproduced when using cokriging with related means.

bias tends to disappear, and the predictions are more consistent, in particular, with a reduction in the amount of predicted soluble grades that are negative or greater than the predicted total grades (Figure 8.18) (Emery, 2012a).

Chapter 9

Conclusions and perspectives

9.1 The importance of indicator variables

Indicator functions have intervened several times in this book: in Chapter 1 to describe the diatreme of El Teniente and the geological facies of the Ministro Hales and Río Blanco-Los Bronces deposits, and in Chapter 2 for the calculation of the partial grade (the product of a grade by an indicator) to create a block model that takes into account the facies specificities. Being able to perform a kriging of such a binary function always surprises at first when compared with the kriging of a quantitative variable like a metal grade. Indeed, a binary function has no nuances, and its spatial variations are abrupt since it passes without transition from 0 to 1, while a grade evolves continuously from one value to another – at least, it is what is expected when calculating the variogram. However, in practice, the experimental variograms of indicator functions usually appear much more continuous and less fluctuating than the experimental variograms of grades, for a very simple reason: for the indicator, the range of possible values is limited to 2 and the variogram cannot have a sill greater than 0.25 (Figure 9.1a), while for a grade, there is a sometimes wide range of possible values and outliers can affect the result of the variogram calculation based on the squared increments of the variable. This stability of indicator variograms is attractive to the geostatistician, who often wishes a 'nice-looking' variogram.

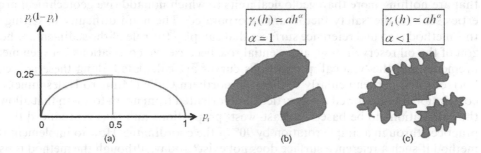

Figure 9.1 Indicator variograms and random sets. (a) The sill of the indicator variogram of set i is the product $p_i(1 - p_i)$, where p_i is the proportion of ones that indicate set i, as compared with the total number of measurements. Depending on p_i, the variogram of the set indicator has a maximum sill of 0.25 when p_i is equal to 0.5. (b) Typical set associated with a variogram behaving, for h close to 0, like h^α with α close to 1. The contours are regular. (c) Typical set associated with a variogram behaving, for h close to 0, like h^α with α close to 0. The contours are irregular.

Another advantage of the indicator function is that its variogram, for a given vector h, is interpreted as the probability that two measurements taken at points separated by h are on both sides of the boundary of the object (called a 'random set' in the geostatistical jargon) represented by the indicator. Therefore, this variogram measures the transition to the geological object in a continuous manner and even characterizes the regularity of its boundary (Lantuéjoul, 2002; Emery and Lantuéjoul, 2011), as illustrated in Figure 9.1b and c. When the variogram of the indicator behaves, at small distances, as a power function h^{α} with α close to 1, the boundary of the object is regular, while for α close to 0, the boundary is diffuse. This concept of recognition of an object by a compact set (the two endpoints of a line segment, or any other topologically closed and bounded set) that moves in the field is one of the foundations of mathematical morphology, a twin discipline of geostatistics created by Georges Matheron and Jean Serra (Matheron and Serra, 2002).

This ability to measure transitions also extends to transitions between different random sets by means of the quotients between the indicator cross- and direct variograms, seen in the analysis of partial grades (Eq. 2.14). This tool has a high potential because it allows determining the propensity of two sets to be in contact or, conversely, not to touch each other (Figure 9.2). It has been used to draw preferential contact diagrams with which mining engineers were able to make important decisions about the exploitation of the deposit (Figure 2.9). This is just one application, there are probably many more, of what will undoubtedly become a cutting-edge geostatistical tool in the future.

9.2 An omnipresent multivariate approach

Cokriging and, more generally, multivariate analyses appear as a second important topic of this book.

From the first chapter, the multivariate approach intervenes. The truncated Gaussian method is used to simulate facies while reproducing their transitions, reflected through indicator cross-variograms. This method is likely to generalize in mining, to model facies or geological units, or analogous objects such as the geotechnical units that are nothing more than geological units to which nonadditive geotechnical properties such as the 'safety factor' are incorporated. The main difficulty in setting up this method is to find reference surfaces that can play the role of the sedimentary horizon of the oil reservoir case, an essential role because the correlations between measurements and the 'vertical' proportions curves are calculated along these reference horizons and perpendicularly to them. In northern Chile (Ministro Hales mine), the correlation lines followed a subvertical fault oriented from north to south that allowed the calculations to be based on 'east–west' proportions and thus reproduced the oil practices through a simple rotation by 90° of the coordinates. How to implement this method if such a reference surface does not exist? Today, although the method is used with different proportion curves at each node of the simulation grid, there is a need to find these reference horizons because they condition the inference of the model parameters, in particular the indicator direct and cross-variograms that heavily depend on the reference system. The structure of the indicator variograms and the continuity of the simulations are often subject to the existence of a surface along which the physical phenomenon measured by the regionalized variable extends, and to the contrast of

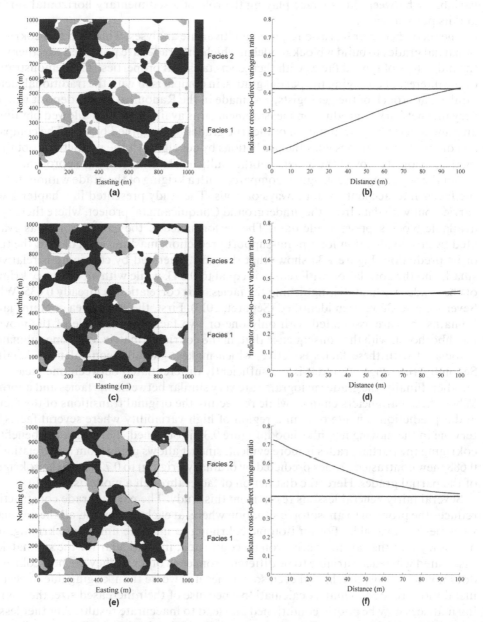

Figure 9.2 Geological facies and quotient $|\gamma_{12}(h)/\gamma_1(h)|$ of their indicator cross-variogram by the direct variogram of the first indicator, depending on the distance h. (a) Facies prone to not having contact: the ratio of the respective indicator variograms (b) is less than the sill $p_2/(1 - p_1)$ (Eq 2.14), indicating the presence of edge effects; (c) facies without preferential relationship (no edge effects): the indicator variogram ratio (d) is identically equal to the sill $p_2/(1 - p_1)$; (e) facies prone to having a lot of contact: the indicator variogram ratio (f) is greater than the sill $p_2/(1 - p_1)$, showing edge effects (Maleki et al., 2017).

variability between this surface playing the role of a sedimentary horizontal surface and its perpendicular.

The second chapter has also required multivariate analyses, which led to a cokriging of partial grades to build a block model in which the grades are predicted together with the indicators of typical facies with their own statistical properties. The demonstration of the interest of cokriging the partial grades, in comparison with the traditional, deterministic approach of the geologists, was made in the Radomiro Tomic deposit where a very fine network of production measurements was available, which allowed obtaining an almost perfect approximation of the true grades of the mined blocks. The comparison of these 'true' grades with the predictions by cokriging or the predictions obtained by the geologists proved the benefits of the multivariate geostatistical approach.

But the advantage of cokriging, compared with a kriging of the grade without taking the facies into account, is not always obvious. The study presented in Chapter 2 was carried out with data from the 'underground Chuquicamata' project, where there is no available block-support grade data. The performance of the calculations was evaluated by cross-validation for a point-support prediction, masking each data at the time of its prediction. Figure 9.3a shows that the gain generated by cokriging is relatively small, and the correlation with reality is equal to 0.643, while with an ordinary kriging of the grades without distinguishing the facies, this correlation is already 0.635. Why? Several causes have been identified (Séguret, 2013). First, the study area contains large domains that are associated with only one or two facies at the scale of the moving neighborhood, with the consequence that, in the cokriging matrix, only one submatrix associated with these facies is active, which makes, globally, equivalent to kriging. Secondly, the range of grades is not sufficiently differentiated between one facies and another. Finally, the grade variograms are very similar between one facies and another. When these parameters change, while respecting the original transitions of the facies, and a prediction is made only in a region of high variability where several facies intervene in the moving neighborhood, Figure 9.3b is obtained. This time, the benefit of cokriging the partial grades is more evident, since it allows going from a correlation of 0.688 (new contrasted grades predicted by ordinary kriging) to 0.727 with the cokriging of the partial grades. Here, the distinction of facies through a model is useful.

Several fairly general lessons result from this study. The partial grade construction reduces the problem to an isotopic situation where, at each data point, all the auxiliary variables are available. For an isotopic cokriging to improve univariate kriging, it is necessary that the variables are strongly contrasted, in particular for the submatrices associated with each variable to be different from each other. If they are not, cokriging does not contribute much, and even sometimes it is worse than kriging due to certain instabilities related to matrix calculations: because of their increased size, the cokriging matrices may be poorly conditioned and lead to inaccurate results. Another lesson is that if, as in the case of poorly contrasted data, the partial grade approach is not effective compared with a kriging that does not consider the facies, is the approach of the geologists, who draw the facies and then predict the grade within each facies, really useful? In other words, at the scale of the production blocks, is it really useful to complicate the task and know that, in the end, grades that vary only a little are averaged between one facies and another? We suspect that, in many cases, these calculations do not bring benefit and a single kriging might be enough. Then it is possible to give a new light to the partial grade approach. While it represents, mathematically, the ideal

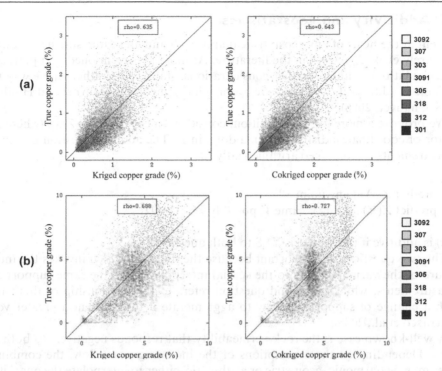

Figure 9.3 Prediction and reality. Scatter plots between predictions (horizontal axes) and reality (vertical axes). On the left, the prediction is a kriging that does not take the facies into account. On the right, the prediction is obtained by cokriging the partial grades. (a) The grades of the deposit are poorly contrasted from one facies to another, and the predictions are made throughout the deposit where there are large areas with only one facies. (b) The grades are strongly contrasted in their means and variograms. The predictions are carried out in a highly variable subdomain where several facies intervene at the scale of the production block.

solution to the problem of predicting the block grades in a multifacies context, it can also be a quick test to know whether or not the grades by facies should be taken into account. If both approaches (direct kriging of the grade or cokriging of the partial grades) provide similar results, then it is likely that the traditional approach of the geologists – separating the facies before predicting the grade of each – is a waste of time; a single kriging of all facies together may be sufficient. In contrast, if the cokriging of the partial grades shows great differences with kriging, then the information of the facies has a significant contribution. In this case, it is important that the geologist distinguishes the facies by either the traditional method or the partial grade one, which can be thought of as another way of drawing the contours of the geological objects.

Other chapters also use multivariate analyses, whether or not they lead to cokriging systems. In Chapter 5, where the grades of drill holes and blast holes are analyzed, cokriging with related means is more precise than traditional cokriging that relates any two variables, since a formal link between the variables has been established using regression formulae. The same is true in Chapter 8 when it comes to predicting total (sampled nonpreferentially) and soluble (subsampled) grades.

9.3 Additivity and geostatistics

Present at the heart of any geostatistical approach and all spatial statistics, additivity is a rarely evoked concept in the literature. At most, it is mentioned as a prerequisite for calculations, without further considerations, stating in a rather vague way that a variable is additive if 'its arithmetic average has a meaning' (Journel and Huijbregts, 1978; Chauvet, 2008).

We propose a more formal definition: consider n measurements $Z(x_i)$ (where x_i is a vector of coordinates) distributed in a domain D. The additive question is: can these measurements be combined arithmetically to

1. predict $Z(x)$ at any point x?
2. predict $Z(V)$ for any volume V, possibly with $V = D$?

$Z(x)$ is additive if the answer is YES to both questions.

The two questions are important because the first one, which invokes the intrinsic nature of the variable, refers to the spatial interpolation on the same support as the measurements, while the second question refers to the relationship of this variable with a change of support, the way to agglomerate it to represent a greater volume (Carrasco et al., 2008).

A well-known case is the rock permeability that responds negatively to both questions. Depending on the assumptions of the incoming water flow, the combination rules may be harmonic, geometric or arithmetic, either to interpolate the quantities or to evaluate them on a larger support. Therefore, a fairly general lesson is that the answer depends on a hypothesis that must be formulated: additivity depends on a context.

Another typical case is the color or shade of gray of a regionalized material, which is a critical issue in the talc industry, for example, where it is important for the management of talc preparations classified by hue and then mixed. It is enough that a small amount of black is mixed with a large amount of white to obtain a very dark gray product, the value of which is considerably lower than pure white product. In this case, the answer to the second question in relation to additivity is NO; it is not possible to arithmetically combine whiteness to express this quantity at the support of a larger volume. A function of the mixture involving the percentage of pure black in the mixture and its complement to 1 of pure white should be experimentally found. The answer to the first question is also negative in the talc deposit of Luzenac, France, that we visited in 2011: when the point x where $Z(x)$ must be evaluated is halfway between two measurements $Z(x_1)$ and $Z(x_2)$, it makes no sense to assign the average of the two values to $Z(x)$, not only due to the nonlinearity of the mixtures but also because in this deposit the talc is in the form of compact objects of an absolute white in direct contact with a very pronounced black substrate. There has been no alteration related to water, and there is no continuous gradation from black to white. Kriging the whiteness on a specific support, which is no more than an optimized average, makes no sense. Here objects should be recognized, with a possible approach being to map or to simulate the indicator of white objects. For whiteness, the two questions involve two different answers, which shows that the concept of additivity is doubly contextual: for the initial assumptions as for permeability and also for how the averages are used, what is expected of them and what they are supposed to represent when interpolated or by change of support.

In geotechnics, intrinsically nonadditive quantities have also been found, such as the rock quality designation (RQD) or the fracture frequency (FF). Without presuming a linearity under a change of support, the directionality of these variables prevents using measurements taken in different directions in the same kriging neighborhood, even if they are corrected for their biases such as FF with the Terzaghi approach. The result remains directional, except if the fracture network is isotropic, that is, if at the scale of the drill-hole samples, the fractures have no tendency to preferentially align along a direction that could change from one place x_1 to another x_2. Depending on the use made of the calculations, it will be possible or not to krige FF, for example, for a mapping, or to assign to FF(x), for x located halfway between x_1 and x_2, the average of FF(x_1) and FF(x_2). Assuming there is no anisotropy and if the mapping intends to correlate measurements of the same nature, a characteristic nature, for example, of the fracturing of the rock, then yes, the operator can consider the variable FF, corrected for its directional bias, as point-support additive. The quality of the map and the value of the safety factors that will be obtained will directly relate to the quality of the hypothesis. It is enough to observe different local anisotropies in x_1 and x_2 and the whole building collapses, both figuratively and literally, in the case of underground works. FF(x_1) and FF(x_2) cannot simply be interpolated to draw a correlation line that indicates a degree of fracturing, because in this case the fracturing depends on the direction along which it is calculated. One must then categorize the samples into several classes of preferential directions of drilling, interpolate the measurements that belong to the same class and obtain as many maps as there are directions, which is not really a complication because an excavation or a drilling to insert a reinforcement is always done in a certain direction, so it is only required to have the corresponding map. The approach is consistent, since it takes into account the nature of the variable that, in reality, is regionalized in a five-dimensional space: the product of the usual three-dimensional space and the angular dimensions of azimuth and dip. In the future, these quantities should be regionalized in a 'superspace' of this type; we will touch this topic again later.

It remains to discuss one last identified family of variables, those for which the additivity is related to the fact that they are the ratios of additive variables: a metal grade when the density of the rock is not constant from one point to another, a problem that arises in particular in the evaluation of iron deposits, where it is preferable to consider the product of the grade by the rock density, additive, instead of the grade itself. Other examples are the reciprocal of a grade, the metallurgical recovery rate and the solubility rate, as already discussed in Chapter 7. It has been seen that, for these variables, the answer to the second question is NO, in theory first and also in practice when the support V is large and/or the dispersion of the point values within this support is important and/or the variance of the numerator, conditionally on the denominator of the quotient, is large, regardless of the value of the quotient. If, at the scale of the production blocks, the variations of these attributes are small, it becomes possible to consider a form of 'local additivity'.

With regard to these magnitudes defined by quotients, the difficulty is centered on the first question: is it permissible to interpolate them spatially, can these calculations be meaningful, are the resulting maps informative? For the FF variable, corrected for its directional bias and in an adirectional context, a method has been proposed in Chapter 6 that provides an unbiased (but not optimal) ratio by dividing the kriging of the numerator and the kriging of the denominator, both krigings having been done with the same variogram model and the same data points. The proof is based on the fact that the quotient of the predictors can be expressed as a linear combination of quotients assigned to weights that add up to 1. Now a question arises: if an ordinary kriging of the quotient had been made directly, there would also be an unbiased result, but it would be different. Therefore, the criterion of unbiasedness of a predictor is not sufficient to choose whether it is adequate or not to interpolate these variables; one must go further and specify what is expected of these predictions, and one returns to the question of the context, how the calculations are used. When a quotient of two krigings is made, two averages are divided, as it is convenient to do when a global recovery rate must be forecast at the scale of a deposit: calculate the quotient of the global averages and not the average of the quotients. When doing this for the interpolation of the recovery rate, the result has the same nature as the data on which it is constructed: an amount of recovered metal divided by an amount of total metal. When a quotient is directly predicted by kriging, a linear combination is produced, which ceases to have the same nature as the data: it is no longer identified with a quotient of metal quantities, and its nature has changed. What can be the consequences if the interpolation does not reproduce the nature of the measurements?

At the time of writing this book in 2020, this question remains largely open and subject to debate. Is it correct to interpolate directly (by kriging or simulation) a nonadditive variable, even if a change of support is not considered? In the case of variables defined by quotients, is it better to predict together the numerator and denominator, which are additive? Pragmatism could provide an answer: the reluctance to interpolate nonadditive variables is probably correct, but the impact on the result can be insignificant in most cases, and what is a precautionary principle can complicate the calculations unnecessarily. Why not try both approaches and compare them by cross-validation?

9.4 From porphyry copper to other deposits

The tools, methods and models presented in this book are not exclusive to porphyry copper deposits and can be used by professionals and specialists working in other types of deposits, for which many problems and solutions can be easily transposed.

However, some applications may require developing new solutions. In particular, one thinks of the following:

- Iron deposits, for which the density of the rock depends on its iron grade: As indicated earlier, the study can be oriented toward additive service variables such as the grades multiplied by the density. Often also, block models must reproduce stoichiometric closure relationships between grades, since their sum must be equal to 100%, a problem that can be addressed through an ad hoc change of variables (Mery et al., 2017).
- Deposits with a narrow geometry, such as a polymetallic vein, a coal seam or a Chilean nitrate deposit called 'caliche': The analysis becomes complicated due to the small size of the deposit along the direction perpendicular to the mineralized envelope. Again, a change in the variables may be appropriate, considering the thickness of the mineralized body and the accumulation (thickness multiplied by the average grade), two additive service variables. This type of deposits can also

be an opportunity to reapply the transitive kriging approach (Matheron, 1967; Renard et al., 2013; also see Section A.5.1 in Appendix);

• The deposits of precious stones (diamonds, emeralds, etc.), for which the use of point processes is an interesting alternative: These processes are defined by a set of random points that simulate the occurrence of the stones, and a set of random 'marks' attached to these points that simulate the properties of the stones (Brown et al., 2008; Lantuéjoul and Millad, 2008). There are also some noteworthy sampling problems when defining the shape, size and spacing of the samples or when taking into account measurements made on samples of different supports (Kleingeld et al., 2005; Duggan et al., 2007; Ferreira and Lantuéjoul, 2007);

• Deposits for which the metal is recovered by pumping, in the same way as hydrocarbons in oil and gas reservoirs. In particular, this is the case with uranium deposits mined by in situ leaching or lithium brine deposits in which the metal is in solution. The concept of classifying the in situ resources no longer makes sense because the slightest pumping action at a given location of the deposit, to a greater or lesser extent, affects the state of the resource at any other location of the deposit. The same goes for the addition of a pumping well that modifies all the predictions of drainable resources. If one also considers a vertical nonstationarity of the mean and the variance of the grades and the need to incorporate duplicates in the model (repeated measurements at the same point in space at different times, with no apparent structure of the time variations), one ends with a chain of treatments that combines geostatistical simulations of the grades, porosity, permeability, upper and lower limits of the deposit with hydrogeological simulations of the pumping, to quantify the uncertainty in the drainable resources (Séguret and Goblet, 2016; Séguret et al., 2017). Interestingly, an analogous problem arises for copper deposits mined by block or panel caving, such as the El Teniente deposit studied in Chapters 1 and 6, for which the in situ resources move during the extraction under the effect of the 'gravitational flow'.

9.5 Perspectives

Throughout our studies, we encountered problems that could be interesting research topics. Here are the most important in our opinion.

9.5.1 Geotechnics and tensor interpolation

Much remains to be done in geotechnics, despite the numerous scientific difficulties that must be overcome. Variables are rarely additive, and they are often not linear when changing the support and are almost always direction-dependent. Interpolating such variables requires interpolating them in space and also in any direction, based on measurements for a few points of space and a few directions. Therefore, it is necessary to develop spatioangular models of covariance (or variogram), and the easiest way is probably to factor such a covariance into the product of a spatial covariance with an angular covariance. This concept of 'separability' has been the subject of many developments in recent years (see, for example, Aston et al., 2016). Since several allowable covariance models in an angular space are also available (see, for example, Melkumyan and Nettleton, 2009), the field seems to be ready to formally address the problem and its natural extent: the spatial prediction of tensors, also useful in structural geology when unfolding geological units to return to a sedimentary context or, more generally, to a context prior to geological deformations, more adapted to geostatistical predictions and simulations. At the time of writing this book (2017–2020), this research work has begun through a joint doctoral thesis between the University of Chile and ParisTech Mines under our direction (Sánchez et al., 2018, 2019).

9.5.2 Geometallurgy

This discipline has grown significantly in recent years. The models of a deposit needed to develop a mining project are no longer limited to the geological and geochemical properties of the rock (facies and grades) but increasingly involve properties of the minerallurgical and metallurgical processes: granulometry, rock density, indices of fragmentation, abrasion, crushing or grinding of the rock, acid consumption (in the case of leaching), mineralogical composition – in particular, clay content – texture, mass recovery, metallurgical recovery, etc. In general, these numerous properties are only known in a very limited number of samples, giving the geological, geochemical and geometallurgical databases a highly heterotopic multivariate character. There are also new challenges, for example, the interpolation of granulometric distributions or of regionalized functions such as flotation kinetics curves, rather than conventional numerical quantities.

9.5.3 Sampling and geostatistics

Unfamiliar in our teams, the sampling theory is a sister discipline of geostatistics. Pierre Gy, its inventor, worked with Georges Matheron, who wrote the preface to his founding (Gy, 1967), as well as a note endorsing the theory developed in this work (Matheron, 1965b). The problem of sampling, in the sense of Pierre Gy, is to explain how to reduce a few kilograms (or hundreds of kilograms) of samples to a few grams, so that the quantities measured in these few grams (for example, the grades) are representative of the kilograms. The techniques to accomplish these objectives consist essentially in mixing the material and finding astute sampling methods. In the end, the material is transformed in such a way that correlations between samples are removed. This probably explains why the two lines of research – sampling and geostatistics – have diverged from the beginning. In essence, geostatistics aims to take advantage of the spatial correlations and the correlations between variables, and not make them disappear.

But there is a link and it is strong, beginning with the variogram and the concept of 'nugget effect', which together constitute the cornerstone of Pierre Gy's theory. Also, depending on the scale considered, there is a sampling in geostatistics, for example, when it is decided to evaluate the economic value of a deposit from a finite number of drill cores that are 'grains' at the scale of the deposit. There is always a sampling as well when an experimental variogram is calculated from a finite number of measurements; this experimental variogram may not represent anything real if some precautions are not taken, as seen in Chapter 8. The consideration of sampling problems in all the geostatistical modeling stages will undoubtedly lead to a closer relation between the two disciplines.

Concepts and practices

Several notions, practical considerations and critical reflection elements related to four stages of geostatistical modeling are collected here:

- exploratory and preparatory data analysis;
- structural analysis;
- kriging and cokriging;
- conditional simulation.

The exploratory and preparatory data analysis aims at providing a comprehensive understanding of the available data and at guiding future choices concerning their modeling.

The cornerstone of the modeling stage, structural analysis or variogram analysis provides tools to quantitatively describe the spatial correlation of regionalized data.

The practice of kriging and cokriging – the predictors mostly used in geostatistics – is often based on the design of a moving neighborhood, which can have catastrophic consequences on the results if a few precautions are not taken. Unfortunately, this problem is too often neglected.

The conditional simulation of categorical (e.g., facies) or quantitative (e.g., grades) variables gives rise to some considerations on the choice of models and simulation algorithms, as well as on the validation of hypotheses and results.

The last section is devoted to two particular classes of geostatistical models: transitive representations and object-based models with, for the latter, examples of applications in mineral processing and geotechnics.

A.1 Exploratory and preparatory data analysis

The following summarizes the fruit of our experience regarding the quality and representativeness of the available information. In particular, geostatisticians should ask themselves about

- the sampling design;
- the accuracy (unbiasedness), precision and support of the measurements;
- the deletion or correction of erroneous data;

- the identification of the relevant variables for the problem posed, often called 'service variables';
- the choice of the domain(s) in which to carry out the study;
- the inference of the distributions of quantitative and categorical variables and of the relationships between these variables.

A.1.1 Data cleaning

Detecting and correcting errors in the database is an essential step. A common practice is to replace an erroneous data with an average value calculated on the nearby samples, which introduces a smooth value of a different nature from the original data and may produce biases in subsequent statistics that aim, among other things, at characterizing the spatial variability. If the correction is not possible, no matter, just delete the erroneous data.

The sources of errors are varied:

- measurement or transcription errors, which result in atypical or inconsistent values; for example, a soluble copper grade greater than the total copper grade is not acceptable for obvious reasons;
- duplication (repetition) of data;
- positioning errors, such as, drill-hole data above the topographic surface;
- inconsistency between different sources of information, such as exploration drill holes, blast holes, interpreted geological model, etc.

Some errors that do not produce any obvious inconsistency may, regrettably, remain undetectable, such as measurement or positioning errors of small amplitude. Apart from a robust checking and validation during the data acquisition, we have no solution for this type of situation.

A.1.2 Data preprocessing

Several operations are grouped under this term, in particular:

- regularization;
- the choice of service variables;
- the truncation of extreme values (capping).

The regularization of the volumetric support of the drill-hole data consists in creating composite samples of (approximately) the same length along the drill holes. The longer these composite samples are, the better the spatial continuity of their values, but the less numerous these samples are.

Choosing the service variables for the study sometimes leads to a change of variables, such as in Chapter 7 where the recovered grade, an additive variable, is considered instead of the recovery rate that is not additive.

Data truncation is a common practice when the distribution of a variable has values that are considered extreme. It consists in lowering (capping) the values that exceed a certain threshold or 'top cut' to allow a more robust calculation of the experimental statistics, such as the mean, variance, covariance or variogram. But this operation induces a bias in the predictions, since the extreme data have been reduced. Rivoirard et al. (2013) propose an elegant solution to correct this bias and to guide the generally arbitrary choice of the truncation threshold.

A.1.3 Partition into geological domains and edge effects analysis

A partition of the deposit into several domains or facies may be required, based on geological, statistical and spatial considerations.

Geological considerations are of fundamental importance, because it is necessary to understand the type of deposit under study and to conceptualize its genesis, as well as the sequence of successive mineralization events. This knowledge allows one to form an idea of the characteristics of the main geological structures or bodies (inclinations, dimensions, ages and temporal relationships, possible presence of faults) and to determine if the information coming from exploration drill holes is adequate or not: types of drilling, inclinations, sampling mesh, core length, geological mapping or logging, types of chemical analyses (Figure A.1).

Statistical and spatial considerations are also essential, because each defined domain must exhibit some 'homogeneity' of the quantitative variables of interest. Several domain definitions can be tested, based on the observed rock types, mineral types, structures and/or alterations and validated by the geological knowledge of the deposit. Analyzing how a variable evolves when getting closer or farther from the boundary that separates two facies (edge effect and contact analysis) makes it possible to decide

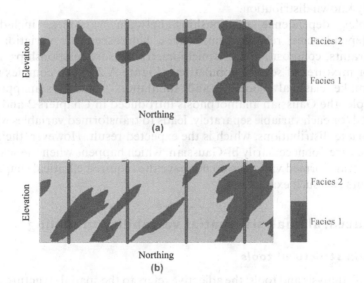

Figure A.1 Geological conceptualization. The spacing and orientation of the drill holes (vertical lines) do not allow determining which of the two conceptualizations is correct, between isotropic facies (a) and anisotropic facies (b). To corroborate or rule out a conceptualization, it is necessary to reduce the spacing between drill holes and/or to add complementary drill holes in oblique directions.

whether two domains in contact should be unified or should remain separated. In order to perform this analysis, several tools have been presented in Chapter 2: quotients of indicator and partial grade cross-variograms and indicator direct variograms, cross-correlograms, lagged scatter plots. An absence of edge effects will guide the modeling toward the formalism of partial grades for the prediction of quantitative variables, or toward a hierarchical approach for their simulation.

In the multivariate case, it is possible that each quantitative variable of interest is associated with a different partition of the deposit, since the same geological controls do not necessarily apply to all the quantitative variables. Mery et al. (2017) present a case study of an iron deposit where the behavior of the iron and silica grades depends mainly on dominant rock type, hematite or itabirite, while granulometry depends on the competence of the dominant rock, its compact or friable character.

A.1.4 Inference of experimental distributions and statistics

This stage is critical if the study is oriented to simulation or to cosimulation, for which a distribution model for every variable separately (simulation) or for all the variables jointly (cosimulation) is required. The main problems are:

- Preferential samplings, which may produce biased experimental distributions that are not representative of the deposit as a whole. A weighting of the data can be used, based on geographical criteria, to eliminate or to mitigate the biases, an aspect already discussed in Chapters 3 and 8.
- Heterotopic samplings, for which some variables are undersampled and have a poorly known distribution.
- Complex dependence relationships between variables, including nonlinear dependencies, related mean values, heteroscedastic variations, inequality constraints, compositional or stoichiometric closure relationships, multimodalities or mixtures of several populations (Figure A.2). Such complex relationships can make classical modeling and simulation techniques inappropriate. For example, the Gaussian anamorphosis introduced in Chapters 2 and 3, when performed for each variable separately, leads to transformed variables with Gaussian univariate distributions, which is the expected result. However, their joint distributions are not necessarily bi-Gaussian, which happens when the scatter plots between transformed variables do not have the required elliptical shape. Figure 7.15 presents two such examples.

A.2 Structural analysis: spatial variability modeling

A.2.1 Main structural tools

'Structural' analysis and tools: the adjective refers to the spatial structure of the regionalized variable (z) and, more specifically, to its continuity and its evolution in space. How to characterize this behavior when the variable is only known at a limited number of samples? Subject to a certain homogeneity that will be further assimilated to a hypothesis of stationarity (see the following), spatial continuity should depend only on the geographical separation between the samples and not on their absolute positions;

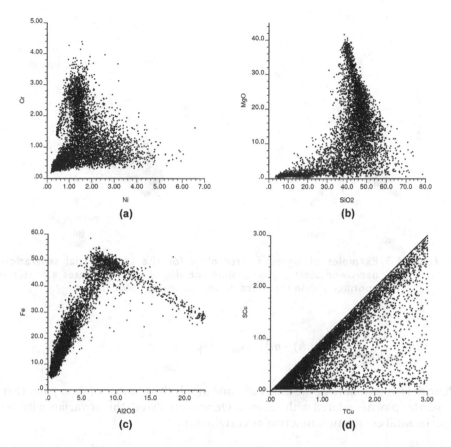

Figure A.2 Heteroscedastic relationships between (a) nickel (Ni) and chromium
(Cr) grades and (b) silica (SiO_2) and magnesia (MgO) grades: the dispersion
of one grade varies according to the value taken by the other grade. (c) Mul-
timodality of the distribution of alumina (Al_2O_3) and iron (Fe) grades: several
groups of points are mixed (increasing part of the scatter plot, upper part,
decreasing part). (d) Inequality between the total copper grade (TCu) and
the soluble copper grade (SCu) and bimodality of their joint distribution:
the points are mainly concentrated along the diagonal or along the horizontal axis, while the area between these two axes appears depopulated.

otherwise, statistical inference would not be possible. A simple way to appreciate this
spatial continuity is to define a separation vector (h) and to compare the $n(h)$ data pairs
$\{z(x_\alpha + h), z(x_\alpha): \alpha = 1 \ldots n(h)\}$ having this separation, via a scatter plot called 'lagged
scatter plot'. For fixed h, the closer the scatter plot to the diagonal, the more similar the
paired data and the more pronounced the spatial continuity. On the contrary, a highly
dispersed scatter plot reflects a weaker spatial continuity (Figure A.3).

Since the lagged scatter plot depends on the chosen separation vector h, measuring
spatial continuity can be summarized, in practice, by quantifying the dispersion of
the scatter plot as a function of h. This can be done in several ways, for instance, by
calculating the experimental centered covariance:

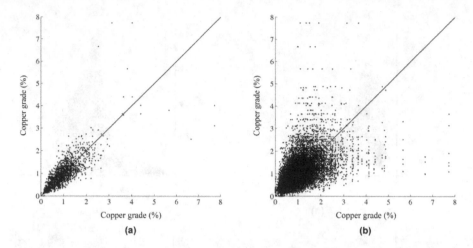

Figure A.3 Examples of lagged scatter plots for the same spatial separation. The narrower scatter plot around the diagonal (a) indicates a better spatial continuity than the more dispersed scatter plot (b).

$$\hat{C}(h) = \frac{1}{n(h)} \sum_{\alpha=1}^{n(h)} \left[z(x_\alpha + h) - \hat{m} \right] \left[z(x_\alpha) - \hat{m} \right],$$ (A.1)

where $\hat{m}$ is an estimate of the mean of z and $n(h)$ is the number of data pairs that form the scatter plot associated with vector h. Once normalized, this covariance leads to the experimental correlation function or correlogram:

$$\hat{\rho}(h) = \frac{\hat{C}(h)}{\hat{C}(0)}.$$ (A.2)

The closer to 1 this correlation is, the more the scatter plot tends to line up around the diagonal. Finally, a third tool is the moment of inertia of the lagged scatter plot, which measures the average quadratic distance between a point of the plot and the diagonal. One of its advantages is not to depend on $\hat{m}$, which also explains the success of this tool defined as the experimental variogram:

$$\hat{\gamma}(h) = \frac{1}{2n(h)} \sum_{\alpha=1}^{n(h)} \left[z(x_\alpha + h) - z(x_\alpha) \right]^2.$$ (A.3)

Figure A.4 shows an example of how these three experimental functions vary with h. The covariance and the correlogram measure the similarity between two data and, in general, decrease when the separation distance increases and vanish when the spatial correlation disappears. The 'range' – the distance from which the covariance and the correlogram vanish – represents the 'distance of influence' of a measurement in its

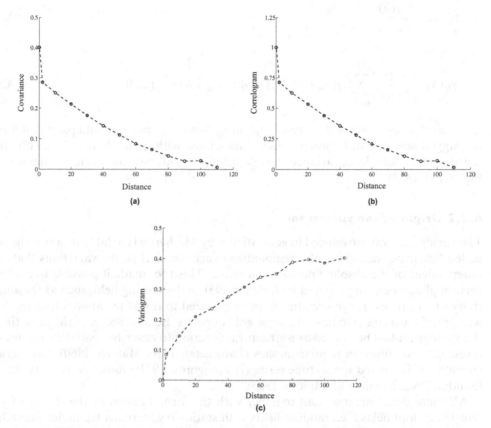

Figure A.4 Three experimental functions that measure the spatial continuity of the same data set. (a) Centered covariance, (b) correlogram, (c) variogram. The calculations were performed in an 'omnidirectional' manner, that is, taking into account the norm of the separation vector *h*, but not its orientation.

environment. On the contrary, the variogram measures the contrast or dissimilarity between two data as a function of the separation between the data positions. It usually appears as a function that increases with the separation distance and stabilizes around a 'sill' when the range is reached. In addition, the behavior near the origin of these tools reflects the regularity of the regionalized variable at small separation distances of the analyzed data pairs, ranging from a parabolic evolution (very strong regularity) to a sharp discontinuity ('nugget effect') (Matheron, 1971).

The aforementioned definitions generalize to the multivariate framework, where one studies N coregionalized variables $z_1 \ldots z_N$. For any pair of indices (i, j) in $[1, \ldots, N]^2$, the experimental cross-covariance, cross-correlogram and cross-variogram are defined as follows:

$$\hat{C}_{ij}(h) = \frac{1}{n_{ij}(h)} \sum_{\alpha=1}^{n(h)} [z_i(x_\alpha + h) - \hat{m}_i][z_j(x_\alpha) - \hat{m}_j] \tag{A.4}$$

$$\hat{\rho}_{ij}(h) = \frac{\hat{C}_{ij}(h)}{\hat{C}_{ij}(0)} \tag{A.5}$$

$$\hat{\gamma}_{ij}(h) = \frac{1}{2n_{ij}(h)} \sum_{\alpha=1}^{n(h)} [z_i(x_\alpha + h) - z_i(x_\alpha)][z_j(x_\alpha + h) - z_j(x_\alpha)] \tag{A.6}$$

where $\hat{m}_i$ is an estimate of the mean of z_i and $n_{ij}(h)$ is the number of data pairs that form the lagged scatter plot between z_i and z_j associated with vector h. If $i=j$, one obtains the 'direct' or 'simple' covariance, correlogram or variogram, as previously defined in Eqs. (A.1)–(A.3).

A.2.2 Origin of the variogram

The variogram was introduced in geostatistics by Matheron (1962a) to describe the so-called 'intrinsic' variations of a regionalized variable, that is, the variations that are independent of the absolute positions in space. This tool made it possible to explain certain phenomena highlighted by Krige in 1951 in the mining field, such as the additivity of variances (Krige's relationship) or the need to resort to random field models with infinite variances to describe some gold deposits. Before 1962 or at the same time, the variogram had been used as a structural function to describe spatial variations of forest data arranged in regular meshes (Langsæter, 1926; Matérn, 1960), time series (Jowett, 1955a, b) and space–time series (Kolmogorov, 1941; Obukhov, 1941, 1949a, b; Gandin, 1963; Monin and Yaglom, 1965).

All these developments went together with the formalization of the theory of intrinsic random fields (i.e., random fields with stationary increments), under the influence of the Russian mathematicians Kolmogorov (1941), Yaglom and Pinsker (1953), Gel'fand (1955), Yaglom (1957, 1958) and Gel'fand and Vilenkin (1964).

A.2.3 Practical calculation

Unless the available data are regularly spaced, for a given separation vector h, the lagged scatter plot will contain few points and will result in a nonrobust calculation of the previous experimental functions. The use of tolerances on the norm and the orientation of vector h is then necessary and requires a certain practical know-how, of which the geostatistician is very proud, but which is likely to raise the hackles of purist statisticians. Using tolerances that are too small leads to less robust experimental functions, because they are calculated with too few data pairs, while too large tolerances tend to smooth out the experimental functions and eliminate their details. Some computer programs also allow weighting the data in the calculations in order to account for irregular sampling designs, a particularly interesting option in the case of a preferential sampling as it allows reducing the biases that such a sampling can produce.

The choice of the directions of calculation must be in accordance with the directional behavior of the regionalized variable. If the variable is isotropic, that is, if it exhibits the same spatial continuity along all the directions of space, the calculations can be 'omnidirectional', regardless of the orientation of the separation vector. In contrast, for an anisotropic variable, it is advisable to first identify the so-called 'main' directions, those for which the spatial correlation is the highest (with the longest correlation range) or the lowest (with the shortest range), before undertaking the calculations along these directions (Figure A.5).

When two variables are studied together, they may not be known at the same positions, in which case the sampling is said to be 'heterotopic' (a neologism forged from the ancient Greek *hetéros*, another, and *topos*, place, the antonym being 'isotopic' formed from the Greek prefix *isos*, equal). This can complicate the calculation of experimental cross-variograms (A.6), which are based on the data pairs for which both variables are known. Highly heterotopic samplings can therefore lead to nonrobust experimental cross-variograms and require the use of cross-covariances (A.4) or cross-correlograms (A.5) instead.

A.2.4 Theoretical covariance, correlogram and variogram

The experimental covariance (A.1), correlogram (A.2) and variogram (A.3) are statistics calculated on a regionalized variable z (lowercase) that, according to the constitutive model of geostatistics, is interpreted as a realization of a random field Z (uppercase). Lowercase and uppercase are writing restrictions introduced by Georges Matheron from the beginning of geostatistics to avoid confusing the model and the regionalized variable, a quite natural temptation when mathematics are confronted with reality. The formalism of random fields makes it possible to replace the experimental averages of Eqs. (A.1)–(A.3) with mathematical expectations and thus define the theoretical version of the three structural functions that, under a hypothesis of stationarity (again), do not depend on the absolute positions of the calculation points (x and $x+h$) but only on their relative position (h):

$$C(h) = E\{[Z(x+h) - m][Z(x) - m]\} \tag{A.7}$$

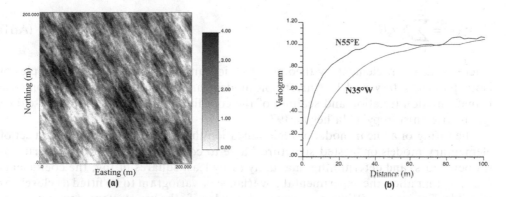

Figure A.5 An anisotropic regionalized variable in the plane (a) and its experimental variogram (b) calculated along the main directions of anisotropy (N35°W and N55°E). The direction parallel to the structures prolongs the spatial correlation (greatest variogram range), while its perpendicular presents the strongest variability (smallest range).

$$\rho(h) = \frac{C(h)}{C(0)} \tag{A.8}$$

$$\gamma(h) = \frac{1}{2} E\left\{[Z(x+h) - Z(x)]^2\right\} = \frac{1}{2} \text{var}\{Z(x+h) - Z(x)\}, \tag{A.9}$$

where m represents the expectation of the random field Z.

The three experimental functions are estimates of these theoretical functions that will serve as a basis for their modeling. The latter is not free (one cannot use just any function for the model) because there are nontrivial mathematical restrictions to ensure the validity and correct representation of a variogram, a covariance or a correlogram, related to their positive definiteness (for the covariance and the correlogram) or conditional negative definiteness (for the variogram).

Consequently, in practice, the modeling is essentially based on linear combinations of elementary functions, also called nested structures, of which one knows that they are theoretically admissible: it is the linear model of regionalization (univariable case) or of coregionalization (multivariate case) (Matheron, 1982; Wackernagel, 2003; Chilès and Delfiner, 2012).

A.2.5 Linear model of regionalization

Consider a set of correlation functions (correlograms) $\rho_1,\ldots,\rho_S$, and decompose the covariance $C(h)$ as follows:

$$C(h) = \sum_{s=1}^{S} b_s\, \rho_s(h) \tag{A.10}$$

where $b_1,\ldots, b_S$ are positive coefficients. This model can be rewritten in terms of the variogram:

$$\gamma(h) = \sum_{s=1}^{S} b_s \gamma_s(h) \tag{A.11}$$

where $\gamma_1,\ldots, \gamma_S$ are elementary theoretical variograms. Figure A.6 shows examples of isotropic elementary models. Anisotropic models can be obtained through a geometric transformation (rotation and scaling) of the coordinates, giving rise to the so-called 'geometric anisotropy' (Matheron, 1971).

The fitting of a linear model of regionalization therefore requires choosing a set of elementary models or 'nested structures' able to correctly represent the experimental behaviors, and determining, manually or by least squares fitting, the coefficients $b_1,\ldots, b_S$ that allow the experimental covariance or variogram to be fitted as closely as possible. For manual fitting, one strategy is to identify the key elements (nugget effect, changes of slope, sills and ranges) of the experimental covariance or variogram along the main directions of anisotropy. These elements then guide the choice of the nested structures, which, when successively added to each other, allow the reproduction of the experimental function. Figure A.7 illustrates this sequence.

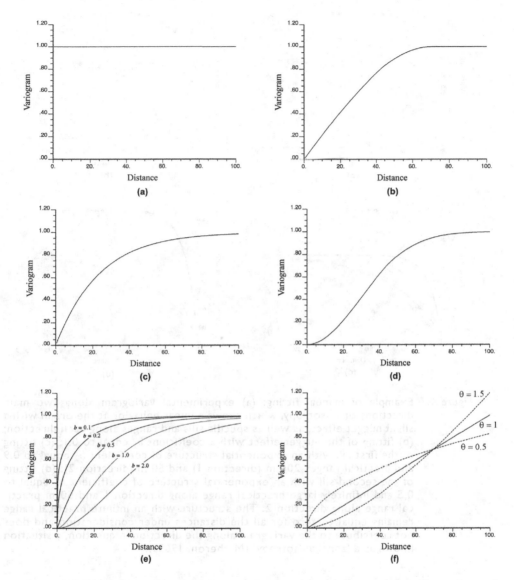

Figure A.6 Examples of elementary variogram models: nugget (a), spherical (b), exponential (c), Gaussian (d), Matérn (e) and power (f). All these models are isotropic and depend on 1 to 3 scalar parameters related to the sill, range or shape of the variogram.

A.2.6 Linear model of coregionalization

The previous formalism extends to the case of several coregionalized variables $z_1, \ldots, z_N$ interpreted as realizations of as many random fields $Z_1, \ldots, Z_N$. For any pair of indices (i, j), the direct (if $i=j$) or cross (if $i \neq j$) covariance, correlogram and variogram between Z_i and Z_j are defined as:

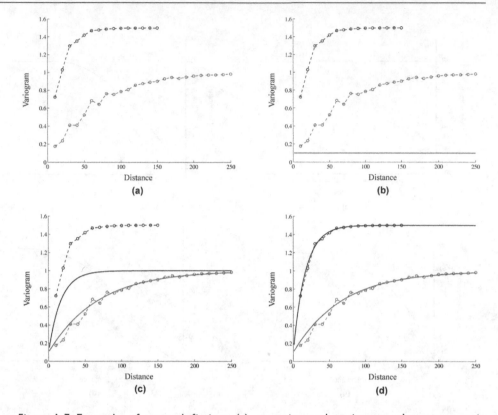

Figure A.7 Example of manual fitting: (a) experimental variogram along two main
directions of anisotropy, which shows a linear behavior at the origin with a
slight nugget effect, as well as specific sills and ranges along each direction;
(b) fitting of the nugget effect with a coefficient b_1 equal to 0.1; (c) fitting
of the first sill with an exponential structure of coefficient b_2 equal to 0.9
and practical ranges 200 m (direction 1) and 50 m (direction 2); (d) fitting
of the second sill with an exponential structure of coefficient b_3 equal to
0.5 and infinitely large practical range along direction 1 and 50 m practical range along direction 2. The structure with an infinite practical range
remains equal to zero for all the distances under consideration and does
not contribute to the variogram along the direction in question, a situation
known as a 'zonal anisotropy' (Matheron, 1965a).

$$C_{ij}(h) = E\left\{[Z_i(x+h) - m_i][Z_j(x) - m_j]\right\} \qquad\qquad (A.12)$$

$$\rho_{ij}(h) = \frac{C_{ij}(h)}{C_{ij}(0)} \qquad\qquad (A.13)$$

$$\gamma_{ij}(h) = \frac{1}{2}E\left\{[Z_i(x+h) - Z_i(x)][Z_j(x+h) - Z_j(x)]\right\}$$
$$= \frac{1}{2}\mathrm{cov}\left\{Z_i(x+h) - Z_i(x), Z_j(x+h) - Z_j(x)\right\} \qquad\qquad (A.14)$$

where m_i represents the expectation of the random field Z_i.

A matrix (bold) notation eases the presentation of this model. For a given separation vector h, consider the matrix $\mathbf{C}(h)$ of direct and cross-covariances, of size $N \times N$, as well as a set of valid correlation functions $\rho_1, ..., \rho_S$ on which one wishes to decompose $\mathbf{C}(h)$. One can write:

$$\mathbf{C}(h) = \begin{pmatrix} C_{11}(h) & \cdots & C_{1N}(h) \\ \vdots & \ddots & \vdots \\ C_{N1}(h) & \cdots & C_{NN}(h) \end{pmatrix} = \sum_{s=1}^{S} \mathbf{B}_s\, \rho_s(h) \tag{A.15}$$

where $\mathbf{B}_1, ..., \mathbf{B}_S$ are real-valued matrices of size $N \times N$, called coregionalization matrices. A sufficient (but not necessary) condition for a model of this type to be valid is that each of the coregionalization matrices be symmetric, positive semidefinite, that is, its eigenvalues be positive or zero. A similar writing for the direct and cross-variograms is:

$$\Gamma(h) = \begin{pmatrix} \gamma_{11}(h) & \cdots & \gamma_{1N}(h) \\ \vdots & \ddots & \vdots \\ \gamma_{N1}(h) & \cdots & \gamma_{NN}(h) \end{pmatrix} = \sum_{s=1}^{S} \mathbf{B}_s\, \gamma_s(h) \tag{A.16}$$

where $\gamma_1, ..., \gamma_S$ are basic theoretical variograms.

The challenge is to identify the adequate nested structures to fit the experimental functions and to determine the coefficients of the coregionalization matrices, such that the restriction of positive definiteness of these matrices is met. If the variables are numerous (N large), the use of automatic or semiautomatic fitting algorithms is very helpful instead of a manual fitting that becomes too laborious (Goulard and Voltz, 1992; Emery, 2010; Desassis and Renard, 2013). Some examples of fitting are shown in Chapters 4, 7 and 8 (Figures 4.3, 7.16 and 8.13).

A.2.7 Stationarity and variogram

Often ignored by practitioners who see it as a complication, at best useless, the concept of stationarity is essential to move from a 'constitutive' model, according to which the regionalized variable is a realization of a random field, to a 'specific' model in which the parameters of this random field are determined by using the measurements taken on this realization. Fundamentally, there is an important problem: the statistics that characterize the random field are based on a repetition over the realizations, while, in our case, the deposit under study is, in essence, unique. The random field is but an abstraction, consisting of an infinite set of realizations that correspond to infinitely many fictitious deposits of which no measurement is available. So how to calculate, for example, the mathematical expectation of the random field $Z(x)$ knowing a finite set of measurements $\{z(x_\alpha): \alpha = 1...n\}$ that sample one of those infinitely many realizations? The first idea that comes to mind is simply to calculate an average of the measurements and deduce without another verification that it is a good approximation of $E[Z(x)]$. But this assumes that this expectation is the same at any point x, so the measurements from different locations of the deposit can be used together for this calculation. There is, therefore, through this practice, an implicit hypothesis of spatial homogeneity, according to which the expectation of Z is constant (hypothesis of stationarity of the mean, i.e., of first order) and coincides with the spatial average of the specific realization z from which the measurements are taken (hypothesis of ergodicity). In the same way, the stationarity and ergodicity hypotheses can be extended to the second-order moments of the random field, which ensure, on the one hand, the existence of a covariance, a correlogram and a variogram that only depend on the separation vector h between the points of calculation (Eqs. A.7 to A.9) and, on the other hand, that these functions can be approximated by their experimental versions (Eqs. A.1–A.3). Moreover, the approximation would be perfect if the measurements covered the entire space.

Are such hypotheses of second-order stationarity and ergodicity always possible? Do they always agree with the measurements?

Ergodicity is often admitted without any further examination, although in practice it can be achieved only if the sampled domain is large enough so that, under the assumption of stationarity, the differences between the experimental spatial statistics and the statistics of the underlying random field are small. Lantuéjoul (1991, 2002) proposes several tests to determine what is meant by a 'large enough' domain.

Concerning stationarity, its validity is debatable when the mean or the variance of the data, calculated locally, varies in space. For example, when analyzing the depth of the El Teniente diatreme in Chapter 1, the stationarity hypothesis, even limited to the first-order moment (mean value), is not satisfactory since an average of the measurements in a neighborhood of a few tens of meters in diameter strongly depends on the position of this neighborhood in the region under study. In fact, formula (A.7) assumes that the global average is representative of each point in space and is (approximately) known, and it is to avoid this pitfall that the experimental variogram is often calculated first. In formulae (A.3) and (A.6), it is not necessary to previously calculate an average, and as such, the calculation of the variogram brings an experimental test of the stationarity of the first two moments of the parent random field. The result generally leads to four different configurations (Figure A.8).

The theoretical variogram (A.9) of a random field is statistically interpreted as half of the variance of its increments. In Figure A.8a, the variance of the increments of the available realization no longer increases after a given distance known as the range and stabilizes around a sill that is approximately equal to the experimental variance of the data. This feature confirms that the mean m and the variance σ^2 of the random field Z associated with this realization do not depend on the spatial position, and consequently, the covariance (A.7) exists and is derived from the variogram by the formula:

$$C(h) = \sigma^2 - \gamma(h). \tag{A.17}$$

The variogram thus allows verifying the adequacy of the second-order stationarity hypothesis with the data. Its sill is nothing else than the variance of the random field:

$$\gamma(h) \xrightarrow[\|h\| \to +\infty]{} \sigma^2 = C(0). \tag{A.18}$$

In Figure A.8b, there is neither a sill nor a range, which disagrees with the assumption of independence in space of the first- and second-order moments of the random field.

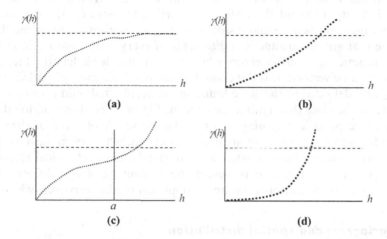

Figure A.8 Four types of behavior of the experimental variogram at large distances: (a) second-order stationarity, (b) intrinsic stationarity, (c) local stationarity at scale 'a', (d) nonstationarity. The dotted horizontal line indicates the experimental variance of the data.

However, the increase is moderate enough to allow the stationarity hypothesis to be placed with a somewhat less demanding level, the so-called 'intrinsic' stationarity in which only the increments of the random field are stationary. This increase rate should not exceed the norm of vector h raised to power 2; otherwise, the experimental calculation (A.3) can no longer be interpreted as a variance of increments where the increments are stationary and have a zero mean (Matheron, 1971). If this is not the case, as in Figure A.8b, we are in a still favorable context where most of the operations carried out in the stationary framework, such as kriging, remain applicable under the condition of modeling the variogram by a function with no sill, for example, the power model that grows indefinitely (Figure A.6f). In this case, the variogram is a generalization of covariance (A.7), which formally does not exist even though its experimental version (A.1) is always calculable from a set of sampling data.

In Figure A.8d, the increase of the experimental variogram is too fast, and one must find something different from the increments of the variable that could be stationary. It may be the residual of the variable when a spatially varying mean – called the 'drift' – responsible for nonstationarity has been removed. This approach is that of simple kriging if the drift is supposed to be known, of 'universal' kriging (Matheron, 1971) or kriging with an 'external drift' (Wackernagel, 2003) if the drift is a weighted sum of basic functions or exhaustively known auxiliary variables, with unknown weights. This 'something different' that is stationary can also be an increment of increments or, more generally, a so-called 'allowable' linear combination that filters all the drifts that are polynomial functions of the coordinates, which leads to the theory of intrinsic random fields of order k or IRF-k. In all these approaches, one has to work with either a variogram of residual, a generalized variogram or a generalized covariance (Matheron, 1971, 1973; Chilès and Delfiner, 2012).

Figure A.8c shows a case in which the variogram, within a certain range of distances, stabilizes around a sill that is lower than the experimental variance of the measurements. Practitioners often refer to this case, qualified as local stationarity or quasistationarity, to avoid the pitfall of resorting to more complex models like the IRF-k, because they believe that it suffices to fit a variogram model at small distances, not to use it at greater distances and to act as if everything were stationary. But this common practice may suffer errors when implemented a bit hastily. First, local stationarity must be verified. It is not enough to simply calculate a global variogram and to apply a model fitted at the scale found in the graph: it should also be verified that this model is the same throughout the deposit. Figure A.8c shows a global curve that may not correspond to anything and only be the result of a regionalization of the high and low grades in different areas of the deposit, which often have very different statistical and spatial characteristics for geological and geochemical reasons. In this case, the global variogram may represent only an average of very different behaviors. By adopting it, the practitioner obtains nonoptimal predictions everywhere.

A.2.8 Variogram and spatial distribution

The variogram, covariance and correlogram are statistics that involve only a pair of random variables $(Z(x), Z(x+h))$ (Eqs. A.7–A.9), so that they are classified among the 'second-order moments' of the random field Z. These tools summarize the main structural properties of this random field (small-scale continuity, anisotropy, correlation range, etc.) but do not describe it completely. While kriging techniques only require these second-order moments, possibly accompanied by the mean value (moment of order 1) for simple kriging, this observation takes its importance in the context of simulation where any indeterminacy on the random field to simulate must be lifted (Figure A.9).

Under assumptions that are not very restrictive in practice, a random field is completely characterized by what geostatisticians call its 'spatial distribution', which corresponds to the set of all the joint probability distributions (also known as finite-dimensional distributions) of vectors $(Z(x_1),..., Z(x_k))$ extracted from this random field, for any choice of the integer k and of the points $x_1,..., x_k$. This is very rich information, even too rich to be inferred from finitely many experimental data, except if one restricts to a particular family of random fields whose spatial distributions depend on few parameters.

The simplest example is that of the Gaussian random field, for which all the finite-dimensional distributions are multinormal or multi-Gaussian. In this particular case, the probability density of the vector $(Z(x_1),..., Z(x_k))$ is:

$$g_{x_1,...x_k}(\mathbf{z}) = \frac{1}{\sqrt{(2\pi)^k \det(\mathbf{C})}} \exp\left\{-\frac{1}{2}(\mathbf{z}-\mathbf{m})^t \mathbf{C}^{-1}(\mathbf{z}-\mathbf{m})\right\} \tag{A.19}$$

where:
  $\mathbf{z} = (z_1,...,z_k)$ is a row vector of k real values;
  $\mathbf{m} = (m_1,...,m_k)$ is the row vector of the mean values at $x_1,..., x_k$;
  $\mathbf{C}$ is the variance–covariance matrix of vector $(Z(x_1),..., Z(x_k))$;
  $\det(.)$ is the determinant and ^{t} is the transpose operator.

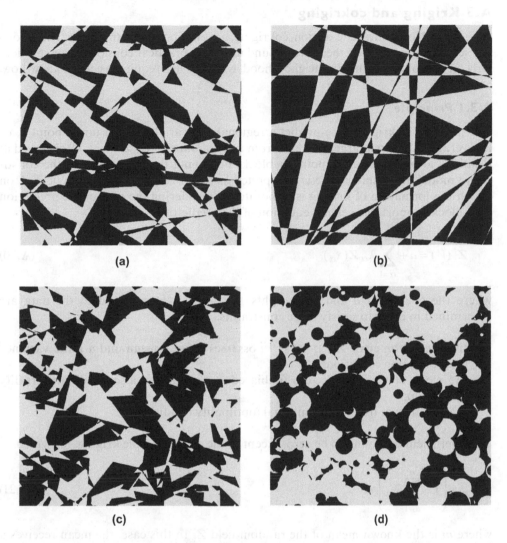

Figure A.9 Realizations of four indicator random fields with the same moments of order 1 (mean equal to 0.5) and order 2 (exponential covariance): Poisson mosaic (a), alternating mosaic (b), STIT mosaic (c), dead leaves mosaic (d) (Emery and Ortiz, 2011). In the context of simulation, it is necessary to go beyond the specification of the covariance or the variogram to distinguish which of these indicator random fields should be modeled and used.

Such a model is attractive. Once the multinormality of the finite-dimensional distributions is admitted, the inference of the random field reduces to the inference of its moments of order 1 (mean) and 2 (covariance or variogram), an easy task in the stationary framework. Second-order stationarity (stationarity of the mean, covariance and variogram) is here the same as the stationarity of the complete spatial distribution (strict stationarity). These remarkable properties largely explain the extensive use of the multi-Gaussian model for simulation.

A.3 Kriging and cokriging

Kriging and its multivariate version, cokriging, are omnipresent in this book. The following sections present their theoretical foundation and practical considerations, particularly on the concept of moving neighborhood, whose design requires a certain know-how.

A.3.1 Principle

The purpose of kriging is to predict a regionalized variable z at a target point x, or its average value on a target block V, from a set of data $\{z(x_\alpha): \alpha = 1,..., n\}$ observed at locations around the target point or block. Following the framework of the constitutive model of geostatistics, z is interpreted as a realization of a random field Z, on which the formalism of kriging is elaborated. Considering the case of the prediction on a block support, the general equation of the predictor is:

$$Z^*(V) = a + \sum_{\alpha=1}^{n} \lambda_\alpha Z(x_\alpha),\tag{A.20}$$

where the coefficient a and the weights $\{\lambda_\alpha: \alpha = 1,..., n\}$ assigned to the data are determined in order to satisfy three constraints:

- the prediction error $Z(V) - Z^*(V)$ possesses a finite mean and a finite variance (authorization condition);
- the mean of the error is zero (unbiasedness condition, also called universality condition);
- the variance of the error is minimal (optimality condition).

The coefficient a is found to be zero, except in the case of simple kriging, for which

$$a = \left(1 - \sum_{a=1}^{n} \lambda_a\right) m,\tag{A.21}$$

where m is the known mean of the random field Z. In this case, the mean receives a weight equal to the complement of the accumulated weight of the data, all the higher as the data are few and/or are located far from the target block. As for the kriging weights $\{\lambda_\alpha: \alpha = 1,..., n\}$, they are obtained by solving the following system of linear equations, written in symbolic form:

$$\begin{pmatrix} C_{\alpha\beta} & F^t \\ F & 0 \end{pmatrix} \begin{pmatrix} \lambda_\alpha \\ \mu \end{pmatrix} = \begin{pmatrix} C_{\alpha V} \\ F_V \end{pmatrix}\tag{A.22}$$

with:

- $C_{\alpha\beta}$ is the variance–covariance matrix of the data vector $(Z(x_1),..., Z(x_n))$.
- $C_{\alpha V}$ is the covariance vector between the data vector and the target value $Z(V)$.

- F and F_V are a matrix and a vector that depend on the authorization and unbiasedness conditions. They are empty in the case of simple kriging (kriging with a known mean). For ordinary kriging (under the hypothesis of second-order stationarity with an unknown mean or of intrinsic stationarity), F is a vector of ones and F_V is equal to 1. Finally, for universal and intrinsic kriging of order k, F and F_V are a matrix and a vector constructed by using the basic functions that model the drift.
- 0 a matrix of zeros, empty for simple kriging and scalar for ordinary kriging.
- λ_α is the vector of kriging weights $(\lambda_1, \ldots, \lambda_n)^t$.
- μ is a vector of Lagrange multipliers, empty for simple kriging and scalar for ordinary kriging.

The covariance function C that allows calculating $C_{\alpha\beta}$ and $C_{\alpha V}$ can/should be replaced by the opposite of the variogram $(-\gamma)$ in the case of ordinary or universal kriging and by a generalized covariance in the case of intrinsic kriging of order k.

In the multivariate isotopic framework, the same Eqs. (A.20)–(A.22) can be written, replacing the scalar quantities (random field Z, mean m, coefficient a, weight λ_α, Lagrange multiplier μ, basic drift functions) by vectors or matrices. In particular, the scalar covariance function C is replaced by a matrix of direct and cross-covariance functions. When the mean values are unknown and unrelated (traditional ordinary cokriging), F is a vector in which each term is an identity matrix and F_V is an identity matrix. In contrast, if the mean values are unknown and equal (cokriging with related means), then F and F_V are a vector of ones and the scalar 1, respectively. In the heterotopic case, the complete system of equations is written first, as if the sampling were isotopic, then the rows and columns of the matrices and vectors of this system corresponding to the missing data are eliminated.

The quality of the prediction is measured by the error variance, known as the 'kriging variance', an ambiguous name that tends to confuse some practitioners and unattended students:

$$\text{var}\{Z(V) - Z^*(V)\} = C_V(0) - \lambda_\alpha^t C_{\alpha V} - \mu^t F_V, \tag{A.23}$$

where C_V is the covariance function regularized on the support V of the target block, with $C_V(0)$ being nothing else than the variance of the variable $Z(V)$ that is targeted for prediction. For ordinary, universal or intrinsic kriging, Eq. (A.23) can/should be rewritten, replacing the covariance function C with the opposite of the variogram or with a generalized covariance function. Besides, in the multivariate framework, the symbolic formula (A.23) is still valid but now provides the variance–covariance matrix of the prediction errors associated with the different variables.

A.3.2 Critical input parameters

Of all the input parameters, four are, according to our experience, crucial for the successful implementation of kriging. The first two are obvious:

- the quality and representativeness of the data – beware of erroneous data or preferential samplings;

- the spatial correlation model represented by direct and, in the multivariate case, cross-covariance functions or variograms.

The third parameter refers to the modeling of the means when simple (co)kriging is used, for which the means must be specified. In practice, a large amount of data is required for this preliminary calculation, which limits the use of a method often based on a contradiction: on the one hand, it is assumed that the mean is known and generally constant; on the other hand, the method is usually set up within a moving neighborhood that integrates, in principle, a 'mild' form of nonstationarity. Another example of modeling of the means has been mentioned several times: cokriging with unknown but related means, a circumstance that was found, for example, with the to-tal copper grades measured in drill holes and blast holes (Chapter 5), or with the total and soluble copper grades measured in drill holes (Chapter 8). The omission of the relationship between the means can lead to a loss of efficiency of cokriging, since an important part of the dependence between the variables stems from this relationship between the means.

A fourth parameter, which has not been mentioned much in this book so far, is the (co)kriging 'neighborhood', a term that refers to the data around the point or block targeted for prediction. The theory is not very prolific in this respect, since the ideal is to use all available information, the famous 'unique neighborhood' that is not always applicable or desirable for several reasons:

- The statistical fluctuations of the experimental variograms increase with the sep-aration distance, which may raise doubts about the quality of the fitted vario-gram model at large distances, distances that are often involved with a unique neighborhood.
- The stationarity hypothesis is often questionable. It is for this reason that, in gen-eral, practitioners prefer ordinary (co)kriging, with unknown mean(s), to simple (co)kriging, because it allows the means to be constant at a local scale but variable at a global scale, without having to specify these local means.
- The amount of available data – sometimes several hundred thousand – can make (co)kriging in a unique neighborhood numerically prohibitive.

Therefore, the practice has ratified the use of a 'moving' or 'local' neighborhood, that is, the selection of a limited subset of the initially available data, those located around and near the target point or block.

A.3.3 Design of a moving neighborhood

In comparison with the unique neighborhood, the use of a moving neighborhood causes a deterioration in the quality of the predictor. Indeed, when a data is removed from the (co)kriging system, the variance of the prediction error increases by an amount equal to the product of two terms (Emery, 2009):

- the square of the weight that is assigned to the removed data;
- the variance of the error made when predicting the removed data on the basis of the other data.

Therefore, it is possible to remove (1) the data that receive a weight close to zero, generally corresponding to the data located very far from the target point or block, and (2) the 'redundant' data, which can be predicted with very little error from other data, without much deterioration of the (co)kriging predictor quality. For instance, in Chapters 4 and 6, the case of data corresponding to grade measurements in drill-hole composites of length 1.5 m has been shown: when selecting one composite in the neighborhood, is it really useful to include the adjacent composites from the same drill hole? The answer will depend on the spatial correlation model, since the 'redundancy' between adjacent data may be small in the presence of a significant nugget effect.

In practice, selecting the best subset of data, the one that leads to the smallest possible variance of error for a predetermined number of data is not simple. Numerous kriging programs allow splitting the space into angular sectors to search for data distributed around the target point, as well as not selecting data that are too close to each other, such as consecutive data of the same drill hole, which are partly redundant. We can only advise the practitioner to test several neighborhoods by means of cross-validation and to choose the neighborhood that has the best performance. In this cross-validation, it is a good practice to mask not only the target sample but also the samples that are very (too) close to the target, such as the samples belonging to the same drill hole; otherwise, the quality of the predictions can be illusory. One variant consists in dividing the data set into K subsets and to predict each subset successively from the other $K-1$ subsets (K-fold cross-validation). The split-sample technique presented in Chapter 2 is nothing more than a particular case of cross-validation with $K=2$ subsets, one of which is used to predict the other.

Within the framework of ordinary kriging with an unknown mean, we recommend taking into account the criteria suggested by Rivoirard (1987) and Vann et al. (2003). First criterion, the weight that is assigned to the mean in the case of a simple kriging should be close to zero, indicating that the selected data are relatively 'abundant'. Second criterion, the slope of the regression of the true values upon the predictions must be close to 1, indicating that the unknown mean is estimated accurately from the selected data and thus avoiding the famous 'conditional bias', a concept that will be detailed in the next subsection. Other parameters can be examined, such as the Lagrange multiplier of the ordinary kriging system, the kriging variance, or the 'kriging efficiency' (Krige, 1997).

In the multivariate framework, the optimal data selection depends on the degree of heterotopy and on the joint spatial correlation of all the variables (Rivoirard, 2004; Subramanyam and Pandalai, 2008). For some coregionalization models where there are proportionality relationships between direct and cross-variograms, the data of the covariates do not provide useful information when they are at the same points as the data of the main variable. In contrast, omitting the data of covariates at points other than the points with data of the main variable can degrade the quality of the predictor and cause conditional biases. Vergara and Emery (2013) show an application for the Ministro Hales (MMH) deposit, where the kriging of the arsenic grade is conditionally biased when a cutoff grade is applied to the kriged copper grade; the conditional bias disappears when the copper and arsenic grades are jointly predicted by cokriging. To avoid omitting any relevant data, Madani and Emery (2019) recommend incorporating the data of the variable of interest located near the target point, as well as the data of the covariates close to the data of the variable of interest or to the target point. In the case of a heterotopic sampling, it is recommended to define a proper neighborhood for each of the variables to predict, even if some coherence is lost between the predictions of the different variables.

A.3.4 Poor neighborhood designs

What are the consequences of a poor design of the (co)kriging neighborhood? By construction, the predictions will remain globally unbiased, but they will become less precise and conditionally – that is, locally – biased:

- less precise because the variance of the error will increase;
- conditionally biased because the regression of the true value upon the predicted value will be different from the true value. Thus, knowing the predicted value, the expectation of the prediction error will be different from zero.

In Chapters 4 and 5, several consequences of conditional biases have been mentioned:

- For long-term planning, conditional bias implies an inaccurate calculation of the grades that will be extracted in the life of the mine, impacting the forecast net present value of the mining project.
- For short-term planning and ore control, conditional bias implies a systematic error in the grade of the ore sent for processing.

Two other consequences are often noticeable:

- Presence of 'artifacts', that is, discontinuities of the predicted values in space that do not correspond to any reality. These artifacts are due to the selection of different subsets of data when predicting values at points close to each other, typically corresponding to adjacent nodes of the grid on which the predictions are made.
- Attenuation of the smoothing effect of (co)kriging. One might think that it is good news to be able to build a predictor that smoothes less and, therefore, allows one to better predict the recoverable resources above a given cutoff grade (see Chapter 3). But less smoothing is also a symptom of less precision and more conditional bias, whose effects can be devastating for short- and long-term mine planning (Krige, 1951, 1997; Journel and Huijbregts, 1978; David, 1988; Montoya et al., 2012; and Chapters 4 and 5).

Another questionable practice is the kriging in several 'passes'. A first pass is made with a very small neighborhood that contains only the data closest to the point or target block; then, a second pass and a third pass are made with larger neighborhoods to predict the values at the points or blocks that could not be predicted in the first pass. Although the motivation of this practice is commendable, since it aims at reducing the smoothing of the predictor, as noted earlier, it causes a loss of precision, an increase of the conditional bias and the emergence of artifacts in the predictions.

A.3.5 Proposals to avoid neighborhood artifacts

Let us mention three approaches to avoid neighborhood artifacts in the kriging results. The first one is the continuous moving neighborhood approximation (Gribov and Krivoruchko, 2004; Rivoirard and Romary, 2011) that penalizes distant data, just as if they were affected by measurement errors whose variance increases as one gets away from the target location, in such a way that their kriging weights become zero when they are beyond a given distance from the target. The second approach is the covariance tapering approximation (Furrer et al., 2006), which consists in multipliying the modeled covariance by a compactly supported covariance (called taper covariance), so that the kriging system becomes sparse, i.e., with many zero entries, and its solution can be computed efficiently with ad hoc algorithms. The last approach is the stochastic partial differential equation (SPDE) approach (Lindgren et al., 2011; Carrizo-Vergara, 2018), which, under some conditions, can yield the kriging problem to a sparse system of linear equations. All three approaches can be used in the presence of large data sets (say, with hundreds of thousands of data) and are likely to gain in importance over the coming years.

A.3.6 Kriging and iterative simulation algorithms

Kriging is one of the components of conditional simulation under the multi-Gaussian model and its variations (truncated Gaussian and plurigaussian models) (Armstrong et al., 2011; Chilès and Delfiner, 2012). Some precautions must be taken with iterative simulation algorithms, for which the values simulated during an iteration are used for the following iterations. Our experience in this area is a high sensitivity of the results to the use of a moving neighborhood, because kriging errors (even small) made by not selecting all the data propagate from one iteration to the next.

Let us mention two particular cases, quite common in the practice of simulation:

1. The sequential Gaussian simulation where, in each iteration, a target point is visited and its value is simulated on the basis of the conditioning data and of the values simulated in the previous iterations. For the propagation of errors to be minimal, the size of the kriging neighborhood and the number of selected data must be very large (Emery and Peláez, 2011), which means that calculation times can become prohibitive. Many other algorithms can be used to simulate Gaussian random fields, among which the turning bands method mentioned several times in this book (Chapters 2, 4, 6 and 7) stands out;
2. The Gibbs sampler, used in truncated Gaussian and plurigaussian simulation to generate Gaussian values at the locations with facies data (Armstrong et al., 2011). Here, the use of a moving neighborhood can be catastrophic, to the point that simulated values diverge if the algorithm does not stop after a finite number of iterations. A variation of the algorithm known under the name of 'Gibbs propagation' avoids this problem and allows the use of a unique neighborhood (Lantuéjoul and Desassis, 2012; Emery et al., 2014).

A.4 Conditional simulation

Referred to throughout this book, conditional simulation aims at creating alternative 'scenarios' or 'outcomes' of one or more variables of interest. They correspond to realizations of the parent random field that, in addition, reproduce the known values at the sampling points. A single simulation gives information about spatial variability, but does not provide a precise prediction and does not allow evaluating the uncertainty in the true values at unsampled points. For this, it is necessary to build numerous simulations – tens, hundreds or thousands. The uncertainty is reflected in the fluctuations observed between one simulation and another at one point in space or jointly at several points.

A.4.1 Facies modeling

The simulation of facies and, more generally, of categorical regionalized variables can be carried out by the truncated Gaussian and plurigaussian models, of which some examples have been seen in Chapters 1 and 2. The plurigaussian model is particularly flexible, since it allows controlling the contact relationships between facies through the definition of a truncation rule, as well as the global or local facies proportions through the definition of the truncation thresholds. In addition, the spatial continuity of the facies is also modeled through the direct and cross-variograms of the Gaussian random fields on which the simulation is based. The first applications of this method in mining have been somewhat late (Skvortsova et al., 2001, 2002; Carrasco et al., 2005) compared with applications to hydrocarbon reservoirs and aquifers, for which this method was initially designed almost fifteen years earlier (Matheron et al., 1987).

The development of new models and the improvement of existing models for facies simulation are current research topics. In particular, the stationarity hypothesis is often questionable when it comes to simulating rock types or mineral zones in a deposit, due to the presence of systematic trends, faults or geological folds that make the Cartesian coordinate system inappropriate.

A.4.2 Grade modeling

The Gaussian random field model (Eq. A.19) is, by far, the most used to simulate, after anamorphosis, the metal grades and, more generally, all kinds of quantitative variables measured on a continuous scale, of which several examples have been given in Chapters 2, 4, 6, and 7. Here are some considerations about this model.

The first one concerns the choice of the simulation algorithm. Users often confuse the *model*, which describes the random field to simulate, and the *algorithm* that describes the mechanism how to perform simulation. Any correctly implemented algorithm should lead to simulations whose properties are those of the model, so the choice of the algorithm is secondary with respect to the choice of the model, although it may have an impact on the calculation time or the memory requirements to construct the simulations. We notice some tendencies that are controversial. Instead of a long argument against a particular algorithm, we will simply mention here two algorithms that have proven their quality, both from the computational point of view and for their flexibility of use: the turning bands algorithm (Matheron, 1973) and one particular case of it, the continuous spectral simulation (Lantuéjoul, 2002; Chilès and Delfiner, 2012; Emery et al., 2016). A summary is presented in Table A.1, which compares the performance of the most widespread algorithms in terms of reproduction of the spatial structure, computational efforts, storage restrictions and conditions of use.

A second consideration refers to the conditioning of the simulations to experimental data. Since most algorithms provide nonconditional simulations, thus reproducing only the spatial structure of the random field, an additional stage is required so that, near the data points, the simulated values get closer to the measured values, until matching them at the data points. This is the conditioning process, which in practice is based on the simple kriging of the error, also called residual, between the measured values and the simulated values at the data points (Chilès and Delfiner, 2012). Two comments on this kriging (or cokriging in the multivariate case):

- In the presence of a large amount of data, kriging must be set up in a moving neighborhood. The definition of a neighborhood that is too small is not only useless – there is no need to fear a smoothing effect in the simulation – but it is also discouraged, since it could introduce an additional undesirable variability

Table A.1 Comparison of the main multi-Gaussian simulation algorithms

Algorithm	Reproduction of the spatial correlation structure	Number of floating point operations required to simulate N points	Need to keep the complete simulation in memory?	Restrictions of use
Cholesky decomposition of the covariance matrix	Exact	$O(N^3)$	Yes	Limited to a few thousand points and few variables
Sequential (unique neighborhood)	Exact	$O(N^3)$	Yes	Limited to a few thousand points and few variables
Sequential (moving neighborhood)	Approximate	$O(N)$	Yes	Limited to few variables
Discrete spectral (FFT)	Exact	$O(N \ln(N))$	Yes	Simulation on a regular grid. Not applicable to all the covariance models
Continuous spectral	Exact	$O(N)$	No	
Turning bands	Exact	$O(N)$	No	

FFT, fast Fourier transform.

(discontinuity between values simulated at adjacent points, due to a change of the neighborhood).

• Although the multi-Gaussian formalism is based on the use of simple kriging, it can be generalized when the mean of the Gaussian variable is uncertain, by substituting ordinary kriging for simple kriging. In practice, this device makes it possible to deal with situations in which the hypothesis of a constant mean value is only acceptable at the scale of the kriging neighborhood (local stationarity).

A.4.3 Validation of the model

Multi-Gaussian simulations have specific characteristics that distinguish them from other random field models. In particular, in addition to a 'diffusive' or 'disseminated' aspect in their spatial structure (like in Figures 2.19, 4.4, 7.17 and A.5), they exhibit the same spatial correlation of the indicators associated with symmetric cutoff values. For example, the indicator associated with the first quartile of the distribution (with three quarters of zeros and a quarter of ones) has the same variogram as the indicator associated with the third quartile (with three quarters of ones and a quarter of zeros).

Since these properties are not always verified with the normal scores transforms of the grade data (after anamorphosis), it is useful to validate them before running the simulation. In addition to the indicator variograms, it has been seen, in the course of the chapters, that the lagged scatter plots are a useful tool to corroborate the multi-Gaussian hypothesis, depending on whether these plots have an elliptical shape or not. Another tool is the variogram of order ω ($\omega > 0$), defined for a stationary random field Y as:

$$\gamma_\omega(h) = \frac{1}{2} E\{|Y(x+h) - Y(x)|^\omega\}. \tag{A.24}$$

Under the multi-Gaussian hypothesis, $\gamma_\omega(h)$ is related to the traditional variogram (of order 2) $\gamma(h)$ by the following relation, where Γ stands for Euler's gamma function:

$$\gamma_\omega(h) = \frac{2^{\omega-1}}{\sqrt{\pi}} \Gamma\left(\frac{\omega+1}{2}\right) [\gamma(h)]^{\omega/2}. \tag{A.25}$$

In a log–log scale, the points representing the variogram of order ω as a function of the variogram of order 2 must be aligned along a straight line of slope $\omega/2$. The validation of the multi-Gaussian hypothesis is, therefore, straightforward. Figure A.10 provides an example for the Gaussian transform of the copper grade data of the Sur-Sur mine at the Río Blanco-Los Bronces deposit, introduced in Chapter 3 (Figure 3.2). Similar validations have been carried out successfully on other porphyry copper deposits, all of a disseminated nature: El Teniente, Chuquicamata, Radomiro Tomic and Gabriela Mistral.

In the multivariate case, the multi-Gaussian hypothesis concerns all the transformed variables jointly. In particular, the scatter plots and lagged scatter plots between different variables must also have an elliptical shape. If this is not the case, an alternative to the anamorphosis of each variable separately is the search for a joint anamorphosis, so that the pairs of transformed variables exhibit elliptic or circular scatter

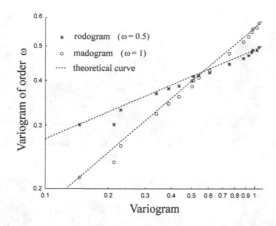

Figure A.10 Río Blanco-Los Bronces deposit. Variograms of order ω = 1 (madogram) and of order ω = 0.5 (rodogram) as a function of the variogram of the normal scores transform of copper grades. On a log–log scale, the experimental points (circles and asterisks, respectively) follow the straight lines (Eq. A.25) expected under a multi-Gaussian model (Emery, 2005a, 2006).

plots. This is certainly a good research topic, since the currently proposed solutions to this problem are difficult to set up in the case of heterotopic samplings or when the number of variables is high.

It is unfortunate that geostatistical modeling software offers too few alternatives to practitioners when the multi-Gaussian hypothesis is not acceptable. As an example of such alternatives, it would be interesting to consider random field models constructed by combinations (compositions, products, nonmonotonic functions, etc.) of multi-Gaussian random fields, which allow obtaining simulations with a nondisseminated aspect or with local anisotropies that evoke, as desired, swirls of smoke, climatic depressions or vortexes of turbulence (Figure A.11).

A.4.4 Validation of the simulation results

As with kriging or cokriging, cross-validation or split-sample methods can be used to check the quality of the constructed simulations. A poor validation could indicate a poor choice of the random field model, or a poor choice of the parameters for setting up the simulation algorithm, in particular the moving neighborhood.

Two tests were presented in Chapter 2:

- a first test of the predictive capacity of simulation, consisting in comparing the true values of the validation data with the average values obtained over a large number of simulations;
- a second test of the ability to quantify uncertainty, based on the experimental determination of probability intervals and on the comparison of the nominal probabilities with the proportions of validation data that effectively belong to these intervals.

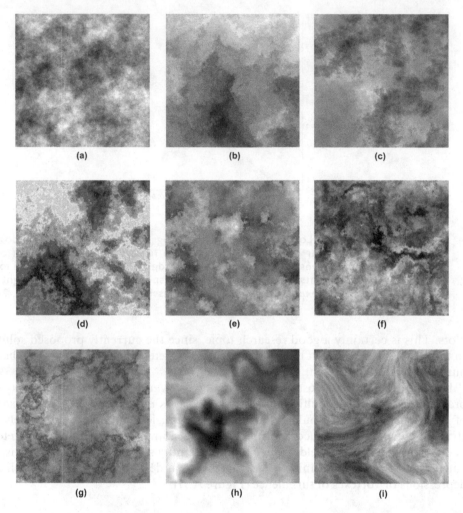

Figure A.11   (a) Simulation of a multi-Gaussian random field, exhibiting a disseminated aspect. (b)–(i) Simulations of random fields defined by combinations of multi-Gaussian random fields, exhibiting a less diffusive structure, patterns of connectivity, as well as spatially varying continuity, smoothness, anisotropy and/or correlation range. All the underlying random field models are, however, stationary and can be simulated conditionally on sampling data by combining Gibbs sampling and multi-Gaussian simulation algorithms, as with the truncated Gaussian and plurigaussian models (Emery, 2007; Emery and Kremer, 2008).

A.4.5 How many simulations?

It is intuitively felt that few simulations cannot adequately quantify the uncertainty in the unknown true value, either at one point, on average in a block or in the entire deposit, because there is always a certain probability for the true value to be out of the range of simulated values. This raises the question of how many simulations are

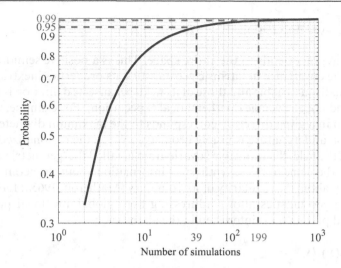

Figure A.12 Probability that the true grade belongs to the range of simulated grades, depending on the number of simulations.

needed to build an interval with a certain level of confidence. Next, a simple criterion is proposed to obtain a tangible answer to this question, which is illustrated by the example of a grade simulation.

Having constructed k simulations, which are statistically indistinguishable because each of them has the same spatial distribution as the reality, the probability that the true grade is greater than the k simulated grades is equal to $1/(k+1)$, an event that becomes increasingly rare as the number of simulations increases. The same applies to the probability that the true grade is less than the k simulated grades, equal to $1/(k+1)$. By difference, the probability that the true grade is between the minimum and maximum simulated grades is $1 - 2/(k+1)$, i.e., $(k-1)/(k+1)$. For example, with 39 simulations, the simulated grades define a 95% probability interval, whereas, with 199 simulations, the probability that the true grade belongs to the range of simulated grades reaches 99% (Figure A.12).

A.5. Transitive representations and object-based models

A.5.1 Transitive geostatistics

Transitive geostatistics is a branch of geostatistics developed by Matheron (1965a, 1971), in which the regionalized variable is studied directly, without interpreting it as a realization of a random field. Therefore, the constitutive model is not the same as the one adopted throughout this book, in particular no hypothesis of stationarity and ergodicity is required.

It is assumed that the regionalized variable vanishes out of a bounded domain of space or 'field', a hypothesis that is verified in practice because a deposit or a mineralized area has always a finite extent. The structural tool is then the 'transitive covariogram':

$$g(h) = \int z(x)z(x+h)\,dx,$$
(A.26)

where the integral extends to the entire space. This is a positive semidefinite function, but unlike the covariance in intrinsic geostatistics, its properties inextricably mix those of the regionalized variable and those of its field, because its definition is based on a sum (integral) and not an average (mathematical expectation). For example, the range of the covariogram in a given direction corresponds to the maximum diameter of the field in this direction and its nature is purely geometric. When the regionalized variable is the indicator of the field, the transitive covariogram is called the 'geometric covariogram', a tool that, as the name implies, synthesizes information about the geometry of the field, in particular, about the regularity of its boundary (Matheron, 1965a; Lantuéjoul, 2002).

Transitive geostatistics allows addressing both global and local prediction problems. Global prediction is interested in the abundance

$$Q = \int z(x)\,dx$$
(A.27)

and allows, under the assumption that the sampling design is regular, stratified random or uniform in the field, constructing an unbiased predictor and calculating the variance of the associated error. The concepts of bias and error variance here stem from the randomization of the sampling mesh and not from the regionalized variable.

Local prediction leads to the so-called 'transitive kriging' (Matheron, 1967), whose equations are formally identical to those of simple or ordinary kriging, except that covariance or the variogram is replaced by the transitive covariogram. This kriging provides a predictor of the regionalized variable in a deterministic context, the notion of error variance being lost. Regrettably less known and less used than its counterpart developed in intrinsic geostatistics, transitive kriging is helpful when the regionalized variable vanishes on the boundary of the field, a situation that is met not only in the evaluation of mineral resources (Renard et al., 2013) but also in the evaluation of fishery resources (Bez et al., 1995; Rivoirard et al., 2000).

A.5.2 Object-based models

Object-based models can be used to represent regionalized quantitative or categorical variables in a construction that considers two ingredients:

- a point process, i.e., an infinite countable set of points randomly distributed in space;
- a family of independent compact (that is, topologically closed and bounded) objects that have the same distribution as a 'typical object', whose shape, dimension, orientation and valuation can be deterministic or random.

An object is placed at each point of the point process. A convention defines the value of the random field at a location where several objects overlap:

- value of the first or the last object, when the point process is not only defined in space but also in time: this is the dead leaves model (Figure A.13a);

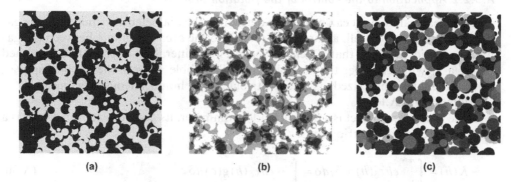

Figure A.13 Object models based on a homogeneous Poisson point process and random discs: (a) dead leaves and (b) dilution using discs valued +1 or −1 with the same probability; (c) Boolean model using discs uniformly valued between −1 and 1.

- sum of the object values: dilution model (Figure A.13b);
- maximum object value: Boolean model (Figure A.13c).

A frequent case of a point process is the homogenous Poisson point process, which corresponds to the concept of randomly and uniformly distributed points in the region under study. This process is characterized by a positive constant θ called intensity, which measures the expected number of points per unit volume.

A.5.2.1 Special case: Boolean random set

When the point process is a homogeneous Poisson process and the objects have binary valuations (0 or 1), the Boolean model provides a binary random field that points to a random set X called Boolean random set. The univariate distribution of this indicator is related to the intensity θ and the expected volume of the typical object A:

$$\text{Prob}\{x \notin X\} = \exp\{-\theta\, E(|A|)\} \tag{A.28}$$

where $|A|$ represents the volume of A. As for the bivariate distribution, it is related to the geometric covariogram $K(h)$ of this typical object:

$$\text{Prob}\{x \notin X, x+h \notin X\} = \exp\{\theta\,[\,K(h) - 2E(|A|)\,]\} \tag{A.29}$$

with $K(h) = E\left\{\int 1_A(x)\,1_A(x+h)\,dx\right\}$.

This model is particularly simple, since it only depends on the definition of a typical object and a scalar parameter (the Poisson intensity θ) related to the number of objects. Simulation conditional on a data set can be performed using iterative acceptance and rejection techniques (Lantuéjoul, 1997, 2002).

A.5.2.2 Application to the control of the flotation process

The Boolean random set can be used to determine the diameter distribution of the gas bubbles in a flotation cell, a parameter related to the area of contact between the gas and the liquid in the cell that affects the recovery of minerals of interest. Photographed inside a viewing chamber that samples the cell (bubble viewer), the bubbles appear, once the photo is binarized, as discs distributed in a homogeneous way and likely to overlap (Figure A.14).

Since the typical object is a disc of random diameter, its geometric covariogram is a mixture of circular covariograms:

$$K(h) = \int_0^{+\infty} circ_\delta(h)\, g(\delta)\, d\delta = \int_h^{+\infty} circ_\delta(h)\, g(\delta)\, d\delta \tag{A.30}$$

with:

- $circ_\delta(h) = \frac{\delta^2}{2}\left\{ \arccos\frac{h}{\delta} - \frac{h}{\delta}\sqrt{1-(\frac{h}{\delta})^2} \right\} 1_{h \le \delta}$: geometric covariogram of the disc of diameter δ;
- $g(\delta)$: probability density of the gas bubble diameters, whose determination is the objective of the study.

The inversion of formula (A.30), based on Leibniz integral rule (differentiation under the integral sign) and Fubini's theorem, gives the complementary distribution function $1 - G$ associated with the density g:

$$1 - G(\delta) = \frac{2}{\pi}\int_\delta^{+\infty} \frac{K''(h)\, dh}{\sqrt{h^2 - \delta^2}}. \tag{A.31}$$

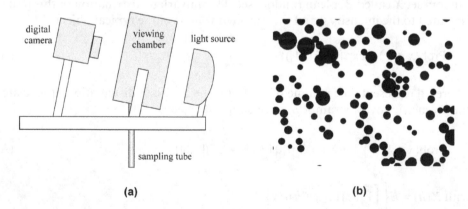

(a)   (b)

Figure A.14 (a) Flotation cell sampling device; (b) binarized photograph of the gas bubbles. The objective is to estimate the distribution of 'the bubble diameters. Image analysis techniques give biased estimates because bubbles can overlap, even be completely hidden, or be cut by the edges of the photograph (Emery et al., 2012; Kracht et al., 2013).

Equation (A.29) allows one to experimentally estimate the covariogram $K(h)$, with the exception of an additive constant and a multiplicative constant (θ). It follows from Eq. (A.31) an unbiased estimate of $\theta(1-G)$, which, once normalized, provides an estimate of the desired complementary distribution function $1-G$. Thus, the structural analysis of the photograph leads to a direct estimation of the bubble diameter distribution, without resorting to more tedious image analysis techniques. Online monitoring of the flotation process is possible, based on the real-time analysis of the photographs taken in the bubble viewer, and can provide feedback to the gas bubble injection system in order to optimize the recovery of the flotation cell.

A.5.2.3 Application to geotechnical modeling

The Boolean model can also be used to describe a network of fractures in a rock mass. For example, fractures can be represented by discs centered at the points of a Poisson process in the three-dimensional space and randomly oriented in this space, a model initially proposed by Baecher et al. (1977) (Figure A.15).

Assuming that the fracture diameters and orientations are mutually independent, the distribution of the diameters of the discs is related to that of their traces observed in a plane via the following formula (Warburton, 1980):

$$f(l) = \frac{1}{\mu} \int_l^{+\infty} \frac{g(\delta)\, d\delta}{\sqrt{\delta^2 - l^2}} \tag{A.32}$$

with:

- $f(l)$: probability density of the fracture trace lengths observed in a plane;
- $g(\delta)$: probability density of the fracture diameters;
- μ: average fracture diameter.

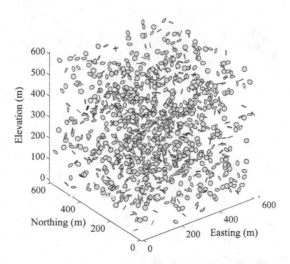

Figure A.15 Boolean model of circular fractures. Each fracture here has a fixed diameter, but its orientation is random (Hekmatnejad et al., 2019).

This formula is still valid if the Poisson intensity, the fracture diameters and fracture orientations are no longer modeled by random variables, but by independent, stationary and ergodic random fields, and if the two-dimensional observation window is large enough for the experimental statistics to be a good approximation of the model statistics. Randomizing the Poisson intensity is equivalent to replacing the Poisson point process with the so-called Cox process (Lantuéjoul, 2002), which allows taking into account the spatial variations in the average number of fractures per unit volume, due to geological and geomechanical factors (stress field direction, rock strength, proximity of faults, etc.).

The inversion of Eq. (A.32) leads to a formula similar to Eq. (A.31) (Hekmatnejad et al., 2018):

$$1 - G(\delta) = \frac{2\mu}{\pi} \int_{l}^{+\infty} \frac{f(l)\,dl}{\sqrt{l^2 - \delta^2}}, \qquad (A.33)$$

expression that allows determining the distribution of the fracture diameters from the trace length distribution, which is experimentally accessible after correcting sampling biases (Baecher and Lanney, 1978, Lantuéjoul et al., 2005).

The simulation of fracture networks represented by Boolean random sets has been studied in particular by Jean-Paul Chilès (Chilès, 1989a, b; Chilès et al., 1992; Chilès and de Marsily, 1993) and remains, to date (2020), an active research topic.

References

Adams, E.P., Hippisley, R.L., 1922, Smithsonian mathematical formulae and tables of elliptic functions. *Smithsonian Miscellaneous Collections* 74(1): 1–314.

Aguirre, A., Emery, X., Goycoolea, M., Moreno, E., 2015. A two-stage stochastic optimization model for open pit mine planning under uncertainty. In: Ardito, J.A., Beniscelli, J., Goycoolea, M., Henríquez, E., Moreno, E., Rojas, S., Villegas, F. (eds.) *Proceedings of the Fourth International Seminar on Mine Planning*. Gecamin Ltda, Santiago, pp. 178–179.

Alegría, A., Emery, X., Lantuéjoul, C., 2020. The turning arcs: a computationally efficient algorithm to simulate isotropic vector-valued Gaussian random fields on the *d*-sphere. *Statistics and Computing* 30(5): 1403–1418.

Armstrong, M., Galli, A.G., Beucher, H., Le Loc'h, G., Renard, D., Doligez, B., Eschard, R., Geffroy, F., 2011. *Plurigaussian Simulations in Geosciences*, 2nd edition. Springer, Berlin.

Armstrong, M., Galli, A.G., Le Loc'h, G., Geffroy, F., Eschard, R., 2003. *Plurigaussian Simulations in Geosciences*, 1st edition. Springer, Berlin.

Aston, J., Pigoli, D., Tavakoli, S., 2016. Tests for separability in nonparametric covariance operators of random surfaces. *The Annals of Applied Statistics* 6: 1906–1948.

Baecher, G.B., Lanney, N.A., 1978. Trace length biases in joint surveys. In: Kim, Y.S. (ed.) *Proceedings of the 19th U.S. Symposium on Rock Mechanics*. American Rock Mechanics Association, New York, pp. 56–65.

Baecher, G.B., Lanney, N.A., Einstein, H.H., 1977. Statistical description of rock properties and sampling. In: Wang, R., Clark, A.B. (eds.) *Proceedings of the 18th U.S. Symposium on Rock Mechanics*. American Rock Mechanics Association, New York, pp. 1–8.

Bai, X., Marcotte, D., Gamache, M., Gregory, D., Lapworth, A., 2018. Automatic generation of feasible mining pushbacks for open pit strategic planning. *Journal of the Southern African Institute of Mining and Metallurgy* 118: 515–530.

Bartlett, R.W., 1997. Metal extraction from ores by heap leaching. *Metallurgical and Materials Transactions B* 28(4): 529–545.

Bartlett, R.W., 1998. *Solution Mining*, 2nd edition. Gordon & Breach Science Publishers, Amsterdam.

Barton, C.C., La Pointe, P.R. (eds.), 1995. *Fractals in the Earth Sciences*. Plenum Press, New York.

Barton, N., Lien, R., Lunde, J., 1974. Engineering classification of rock mass for the design of tunnel support. *Rock Mechanics* 6(4): 189–226.

Beniscelli, J., 2011. Geometallurgy – fifteen years of developments in Codelco: Pedro Carrasco contributions. In: Dominy, S. (ed.) *Proceedings of the First AusIMM International Geometallurgy Conference*. Australasian Institute of Mining and Metallurgy, Carlton, pp. 3–8.

Beucher, H., Chilès, J.P., Rivoirard, J., Séguret, S.A., 2008. Geostatistical consulting, tutoring, development and implementation contract – Global Report of Year 3 (2007–2008). Contrat ARMINES/CODELCO Chili. Mines ParisTech. R080824SSEG, 75 p.

Beucher, H., Galli, A., Le Loc'h, G., Ravenne, C., Heresim Group, 1993. Including a regional trend in reservoir modelling using the truncated Gaussian method. In: Soares, A. (ed.) *Geostatistics Tróia'92*. Kluwer Academic Publishers, Dordrecht, pp. 555–566.

Bez, N., Rivoirard, J., Poulard, J.C., 1995. Approche transitive et densité de poissons. In: de Fouquet, C. (ed.) Cahiers de Géostatistique, Fascicule 5. École des Mines de Paris, Fontainebleau, pp. 161–177.

Brown, G., Ferreira, J., Lantuéjoul, C., 2008. Conditional simulation of a Cox process using multiple sample supports. In: Ortiz, J.M., Emery, X. (eds.) *Proceedings of the 8th International Geostatistics Congress*. Gecamin Ltda, Santiago, pp. 459–468.

Brzovic, A., 2009. Rock mass strength and seismicity during caving propagation at the El Teniente Mine, Chile. In: Tan, C.A. (ed.) *7th International Symposium on Rockbursts and Seismicity in Mines (RaSiM7)*. Rinton Press, New York, pp. 838–852.

Brzovic, A., Villaescusa, E., 2007. Rock mass characterization and assessment of block-forming geological discontinuities during caving of primary copper ore at the El Teniente mine, Chile. *International Journal of Rock Mechanics and Mining Sciences & Geomechanics Abstracts* 44: 565–583.

Caccetta, L., 2007. Application of optimisation techniques in open pit mining. In: Weintraub, A., Romero, C., Bjørndal, T., Epstein, R., Miranda, J. (eds.) *Handbook of Operations Research in Natural Resources*. International Series in Operations Research & Management Science vol. 99. Springer, Boston, MA, pp. 547–559.

Candelario, L., 2010. Analyse géostatistique et estimation d'un gisement polymétallique Pasco – Pérou. CFSG thesis. École des Mines de Paris, Fontainebleau.

Carlson, T.R., Erickson, J.D., O' Brian, D.T., Pana, M.T., 1966. Computer techniques in mine planning. *Mining Engineering* 18(5): 53–56.

Carrasco, P., 2010. Nugget effect, artificial or natural? *Journal of the Southern African Institute of Mining and Metallurgy* 110: 299–305.

Carrasco, P., Chilès, J.P., Séguret, S.A., 2008. Additivity, metallurgical recovery, and grade. In: Ortiz, J.M., Emery, X. (eds.) *Proceedings of the Eighth International Geostatistics Congress*. Gecamin Ltda, Santiago, pp. 237–246.

Carrasco, P., Carrasco, P., Ibarra, F., Le Loc'h, G., Rojas, R., Séguret, S.A., 2005. Application of the truncated Gaussian simulation method to the Mansa Mina deposit at Codelco Norte, Chile. In: Deraisme, J., Le Loc'h, G. (eds.) *Workshop 7 Mining Geostatistics from Exploration to Production,* 67th European Association of Geoscientists and Engineers Conference and Technical Exhibition, EAGE 2005. European Association of Geoscientists and Engineers, Madrid, pp. 15–18.

Carrasco, P., Ibarra, F., Rojas, R., Le Loc'h, G., Séguret, S.A., 2007. Application of the truncated Gaussian simulation method to a porphyry copper deposit. In: Magri, E. (ed.) *Proceedings of the 33rd International Symposium on Applications of Computers and Operations Research in the Mineral Industry*. Gecamin Ltda, Santiago, pp. 31–39.

Carrizo-Vergara, R., 2018. Development of geostatistical models using stochastic partial differential equations. PhD thesis, Mines ParisTech & PSL Research University.

Chauvet, P., 2008. *Aide-Mémoire de Géostatistique Linéaire*. Les Presses de l'École des Mines de Paris, Paris.

Chilès, J.P., 1989a. Fractal and geostatistical methods for modeling a fracture network. *Mathematical Geology* 20(1): 33–40.

Chilès, J.P., 1989b. Modélisation géostatistique de réseaux de fractures. In: Armstrong, M. (ed.) *Geostatistics*. Kluwer Academic Publishers, Dordrecht, pp. 57–76.

Chilès, J.P., Delfiner, P., 2012. *Geostatistics: Modeling Spatial Uncertainty*. Wiley, New York.

Chilès, J.P., Guérin, F., Billaux, D., 1992. 3D stochastic simulation of fracture network and flow at Stripa conditioned on observed fractures and calibrated on measured flow rates. In: Tillerson, J.R., Wawersik, W.R. (eds.) *Rock Mechanics*. Balkema, Rotterdam, pp. 533–542.

Chilès, J.P., de Marsily, G., 1993. Stochastic models of fracture systems and their use in flow and transport modeling. In: Bear, J., Tsang, C.F., de Marsily, G. (eds.) *Flow and Contaminant Transport in Fractured Rock*. Academic Press, San Diego, CA, pp. 169–236.

Chilès, J.P., Séguret, S.A., 2009. Geostatistical study of Al-Khabra phosphate ore body. Rapport Technique. Centre de Géostatistique, École des Mines de Paris, Fontainebleau.

Chilès, J.P., Wackernagel, H., Beucher, H., Lantuéjoul, C., Elion, P., 2008. Estimating fracture density from a linear or areal survey. In: Ortiz, J.M., Emery, X. (eds.) *Proceedings of the 8th International Geostatistics Congress*. Gecamin Ltda, Santiago, pp. 535–544.

Coléou, T., 1987. Paramétrage technique des réserves et optimisation d'un projet minier, Thèse de docteur-ingénieur en sciences et techniques minières, Centre de géostatistique, École des Mines de Paris, Fontainebleau, France.

Cuadra, P., Rojas, G., 2001. Oxide mineralization at the Radomiro Tomic porphyry copper deposit, northern Chile. *Economic Geology* 96(2): 387–400.

Dagbert, M., Maréchal, A., 2008. The Boletín de Geoestadística (1972–73): a first journal devoted to geostatistics. In: Ortiz, J.M., Emery, X. (eds.) *Proceedings of the Eighth International Geostatistics Congress*. Gecamin Ltda, Santiago, pp. 3–9.

David, M., 1988. *Handbook of Applied Advanced Geostatistical Ore Reserve Estimation*. Elsevier, Amsterdam.

Deere, D.U., Hendron, A.J., Patton, J.F.D., Cording, E.J., 1967. Design of surface and near-surface construction in rock. In: Fairhust, C. (ed.) *Failure and Breakage of Rock, Proceedings of 8th US Symposium Rock Mechanics*. Society of Mining Engineers, American Institute of Mining, Metallurgical and Petroleum Engineers, New York, pp. 237–302.

Deraisme, J., Field, M., 2006. Geostatistical simulations of kimberlite orebodies: application to sampling optimization. In: Dominy, S. (ed.) *Proceedings of the 6th International Mining Geology Conference, Darwin*. Australasian Institute of Mining and Metallurgy, Melbourne, pp. 193–203.

Deraisme, J., Strydom, M., 2009. Estimation of iron resources integrating diamond and percussion drillholes. In: *Proceedings of the 34th International Symposium on Application of Computers and Operations Research in the Mineral Industry*. Canadian Institute of Mining, Matallurgy and Petroleum, Vancouver, pp. 49–58.

Dershowitz, W.S., Herda, H.H., 1992. Interpretation of fracture spacing and intensity. In: Tillerson, J.R., Wawersik, W.R. (eds.), *Rock Mechanics*. A.A. Balkema, Rotterdam, pp. 757–766.

Desassis, N., Renard, D., 2013. Automatic variogram modeling by iterative least squares: univariate and multivariate cases. *Mathematical Geosciences* 45(4): 453–470.

Desnoyers, Y., Dogny, S., 2014. Geostatistical deconvolution with non destructive measurements for radiological characterization of contaminated facilities. In: Jeannée, N., Romary, T. (eds.) *Geostatistics for Environmental Applications: GeoEnv 2014 Book of Abstracts*. Presses des Mines, Paris, pp. 50–51.

Dimitrakopoulos, R. (ed.), 2007. *Orebody Modelling and Strategic Mine Planning*. Australasian Institute of Mining and Metallurgy, Carlton South.

Dimitrakopoulos, R. (ed.), 2011. *Advances in Orebody Modelling and Strategic Mine Planning*. Australasian Institute of Mining and Metallurgy, Carlton South.

Dimitrakopoulos, R. (ed.), 2018. *Advances in Applied Strategic Mine Planning*. Springer, Cham.

Dimitrakopoulos, R., Ramazan, S., 2008. Stochastic integer programming for optimizing long term production schedules of open pit mines. methods, application and value of stochastic solutions. *Transactions of the Institutions of Mining and Metallurgy: Section A Mining Technology* 117(4): 155–160.

Duggan, S., Lantuéjoul, C., Prins, C., 2007. Towards an optimum sample support for estimating the diamond density of placer deposits. In: Costa, J.F., Koppe, J. (eds.) *Proceedings of the 3rd World Conference on Sampling and Blending*. Publication Series N° 1/2007, Fundaçao Luiz Englert, Porto Alegre, pp. 126–137.

Emery, X., 2005a. Variograms of order ω: a tool to validate a bivariate distribution model. *Mathematical Geology* 37(2): 163–181.

Emery, X., 2005b. Simple and ordinary multigaussian kriging for estimating recoverable reserves. *Mathematical Geology* 37(3): 295–319.

Emery, X., 2006. Two ordinary kriging approaches to predicting block grade distributions. *Mathematical Geology* 38(7): 801–819.

Emery, X., 2007. Using the Gibbs sampler for conditional simulation of Gaussian-based random fields. *Computers & Geosciences* 33(4): 522–537.

Emery, X., 2009. The kriging update equations and their application to the selection of neighboring data. *Computational Geosciences* 13(3): 269–280.

Emery, X., 2010. Iterative algorithms for fitting a linear model of coregionalization. *Computers & Geosciences* 36(9): 1150–1160.

Emery, X., 2012a. Cokriging random fields with means related by known linear combinations. *Computers & Geosciences* 38(1): 136–144.

Emery, X., 2012b. Co-simulating total and soluble copper grades in an oxide ore deposit. *Mathematical Geosciences* 44(1): 27–46.

Emery, X., Arroyo, D., Peláez, M., 2014. Simulating large Gaussian random vectors subject to inequality constraints by Gibbs sampling. *Mathematical Geosciences* 46(3): 265–283.

Emery, X., Arroyo, D., Porcu, E., 2016. An improved spectral turning-bands algorithm for simulating stationary vector Gaussian random fields. *Stochastic Environmental Research and Risk Assessment* 30(7): 1863–1873.

Emery, X., Carrasco, P., Ortiz, J., 2004. Modelamiento geoestadístico de la razón de solubilidad en un yacimiento de oxidados de cobre. In: Magri, E.J., Ortiz, J.M., Knights, P., Vera, M., Henríquez, F., Barahona, C. (eds.) *1st International Conference on Mining Innovation MININ 2004*. Gecamin Ltda, Santiago, pp. 226–236.

Emery, X., Cornejo, J., 2010. Truncated Gaussian simulation of discrete-valued, ordinal coregionalized variables. *Computers & Geosciences* 36(10): 1325–1338.

Emery, X., González, K., 2007. Incorporating the uncertainty in geological boundaries into mineral resources evaluation. *Journal of the Geological Society of India* 69(1): 29–38.

Emery, X., Kracht, W., Egaña, A., Garrido, A., 2012. Using two-point set statistics to estimate the diameter distribution in Boolean models with circular grains. *Mathematical Geosciences* 44(7): 805–822.

Emery, X., Kremer, F., 2008. A survey of random field models for simulating mineral grades. In: Ortiz, J.M., Emery, X. (eds.) *Proceedings of the Eighth International Geostatistics Congress*. Gecamin Ltda, Santiago, pp. 157–166.

Emery, X., Lantuéjoul, C., 2006. TBSIM: a computer program for conditional simulation of three-dimensional Gaussian random fields via the turning bands method. *Computers & Geosciences* 32(10): 1615–1628.

Emery, X., Lantuéjoul, C., 2011. Geometric covariograms, indicator variograms and boundaries of planar closed sets. *Mathematical Geosciences* 43(8): 905–927.

Emery, X., Maleki, M., 2019. Geostatistics in the presence of geological boundaries: application to mineral resources modeling. *Ore Geology Reviews* 114: 103124.

Emery, X., Ortiz, J.M., 2007. Weighted sample variograms as a tool to better assess the spatial variability of soil properties. *Geoderma* 140(1–2): 81–89.

Emery, X., Ortiz, J.M., 2011. A comparison of random field models beyond bivariate distributions. *Mathematical Geosciences* 43(2): 183–202.

Emery, X., Ortiz, J.M., Rodríguez, J.J., 2006. Quantifying uncertainty in mineral resources by use of classification schemes and conditional simulations. *Mathematical Geology* 38(4): 445–464.

Emery, X., Peláez, M., 2011. Assessing the accuracy of sequential Gaussian simulation and cosimulation. *Computational Geosciences* 15(4): 673–689.

Emery, X., Porcu, E., 2019. Simulating isotropic vector-valued Gaussian random fields on the sphere through finite harmonics approximations. *Stochastic Environmental Research and Risk Assessment* 33(8–9): 1659–1667.

Emery, X., Séguret, S.A., 2020. *Geoestadística de Yacimientos de Cobre Chilenos – 35 Años de Investigación Aplicada*. Caligrama, Seville.

Emery, X., Silva, D.A., 2009. Conditional co-simulation of continuous and categorical variables for geostatistical applications. *Computers & Geosciences* 35(6): 1234–1246.

Emery, X., Soto-Torres, J.F., 2005. Models for support and information effects: a comparative study. *Mathematical Geology* 37(1): 49–68.

Espinoza, D., Goycoolea, M., Moreno, E., Muñoz, G., Queyranne, M., 2013a. Open pit mine scheduling under uncertainty: a robust approach. In: Costa, J.F., Koppe, J., Peroni, R. (eds.) *Proceedings of the 36th International Symposium on Applications of Computers and Operations Research in the Mineral Industry*. Fundação Luiz Englert, Porto Alegre, pp. 433–444.

Espinoza, D., Lagos, G., Moreno, E., Vielma, J.P., 2013b. Risk averse approaches in open-pit production planning under ore grade uncertainty: a ultimate pit study. In: Costa, J.F., Koppe, J., Peroni, R. (eds.) *Proceedings of the 36th International Symposium on Applications of Computers and Operations Research in the Mineral Industry*. Fundação Luiz Englert, Porto Alegre, pp. 492–501.

Ferreira, J., Lantuéjoul, C., 2007. Compensation for sample mass irregularities in core sampling for diamonds and its impact on grade and variography. In: Costa, J.F., Koppe, J. (eds.) *Proceedings of the 3rd World Conference on Sampling and Blending*. Publication Series N° 1/2007, Fundaçao Luiz Englert, Porto Alegre, pp. 3–15.

Fontaine, L., Beucher, H., 2006. Simulation of the Muyumkum uranium roll front deposit by using truncated plurigaussian method. In: Dominy, S. (ed.) *Proceedings of the 6th International Mining Geology Conference, Darwin*. Australasian Institute of Mining and Metallurgy, Melbourne, pp. 205–215.

Fouedjio, F., 2016. Space deformation non-stationary geostatistical approach for prediction of geological objects: case study at El Teniente mine (Chile). *Natural Resources Research* 25(3): 283–296.

Fouedjio, F., Séguret, S.A., 2016. Predictive geological mapping using closed-form non-stationary covariance functions with locally varying anisotropy: case study at El Teniente mine (Chile). *Natural Resources Research* 25(4): 431–443.

François-Bongarçon, D., 1978. Le paramétrage des contours optimaux d'une exploitation à ciel ouvert. Thèse de docteur-ingénieur. Institut National Polytechnique de Lorraine, Nancy.

Fuerstenau, M.C, Jameson, G., Yoon, R.H. (eds.), 2007. *Froth Flotation: A Century of Innovation*. Society for Mining, Metallurgy, and Exploration, Littleton, CO.

Furrer, R., Genton, M.G., Nychta, D.W., 2006, Covariance tapering for interpolation of large spatial datasets. *Journal of Computational and Graphical Statistics* 15(3): 502–523.

Gandin, L.S., 1963. *Obektivnyi analiz meteorologicheskikh polei*. Gidrometeologicheskoe Izdatel'stvo, Leningrad.

Gel'fand, I.M., 1955. Generalized random processes. *Doklady Akademii Nauk SSSR* 100: 853–856.

Gel'fand, I.M., Vilenkin, N.Y., 1964. *Generalized Functions*. Applications of Harmonic Analysis, Vol. 4. Academic Press, New York.

González, R., Brzovic, A., 2015. Características de las fallas con potencial de generar estallidos de rocas, en mina El Teniente. In: Aracena, I., Freire, R., Ibarra, F., Menzies, A. (eds.) *Proceedings of the Fourth International Seminar on Geology for the Mining Industry*. Gecamin Ltda, Santiago, pp. 73.

Goulard, M., Voltz, M., 1992. Linear coregionalization model: tools for estimation and choice of cross-variogram matrix. *Mathematical Geology* 24(3): 269–286.

Gribov, A., Krivoruchko, K., 2004. Geostatistical mapping with continuous moving neighborhood. *Mathematical Geology* 36(2): 267–281.

Gy, P., 1954. Erreur commise dans le prélèvement d'un échantillon sur un lot de minerai. *Revue de l'Industrie Minérale* 36: 311–345.

Gy, P., 1956. Poids à donner à un échantillon. Abaques d'échantillonnage. *Revue de l'Industrie Minérale* 38(636): 53–99.

Gy, P., 1967. *L'Échantillonnage des Minerais en Vrac*. Mémoires du Bureau de Recherches Géologiques et Minières, n°56. Éditions BRGM, Paris.

Hekmatnejad, A., Emery, X., Brzovic, A., Schachter, P., Vallejos, J.A., 2017. Spatial modeling of discontinuity intensity from borehole observations at El Teniente mine, Chile. *Engineering Geology* 228: 97–106.

Hekmatnejad, A., Emery, X., Elmo, D., 2019. A geostatistical approach to estimating the parameters of a 3D Cox-Boolean discrete fracture network from 1D and 2D sampling observations. *International Journal of Rock Mechanics and Mining Sciences* 113: 183–190.

Hekmatnejad, A., Emery, X., Vallejos, J.A., 2018. Robust estimation of the fracture diameter distribution from the true trace length distribution in the Poisson-disc discrete fracture network model. *Computers and Geotechnics* 95: 137–146.

Herrera, A., 2013. Using blast and brill holes together for grade estimation – application on a copper deposit. CFSG thesis, École des Mines de Paris, Fontainebleau, France – N° C130629AHER.

Hoal, K.O., Woodhead, J.D., Smith, K.S., 2013. The importance of mineralogical input into geometallurgy programs. In: Dominy, S. (ed.) *Proceedings of the Second AusIMM International Geometallurgy Conference*. Australasian Institute of Mining and Metallurgy, Carlton, pp. 17–25.

Hochbaum, D.S., Chen, A., 2000. Performance analysis and best implementations of old and new algorithms for the open-pit mining problem. *Operations Research* 48(6): 894–914.

Huijbregts, C., 1971. *Reconstitution du variogramme ponctuel à partir d'un variogramme expérimental régularisé*. Note N-244. Centre de Géostatistique, École des Mines de Paris, Fontainebleau.

Hustrulid, W.A., Bullock, R.L. (eds.), 2001. *Underground Mining Methods: Engineering Fundamentals and International Case Studies*. Society for Mining, Metallurgy, and Exploration, Littleton, CO.

Hustrulid, W.A., Kuchta, M., Martin, R.K, 2013. *Open Pit Mine Planning and Design*. CRC Press, Bosa Roca, FL.

Jaeger, J.C., Cook, N.G.W., 1969. *Fundamentals of Rocks Mechanics*. Methuen, London.

Jeulin, D., Renard, D., 1992. Practical limits of the deconvolution of images by kriging. *Microscopy Microanalysis Microstructures* 3(4): 333–361.

Jones, R.H., 1963. Stochastic processes on a sphere. *Annals of the Institute of Mathematical Statistics* 34: 213–218.

Journel, A.G., 1977. *Géostatistique Minière*. Cours L-476. Centre de Géostatistique, École des Mines de Paris, Fontainebleau.

Journel, A.G., Huijbregts, C.J., 1978. *Mining Geostatistics*. Academic Press, London.

Journel, A.G., Segovia, R., 1974. Improving block estimations at El Teniente copper mine (Chile). Rapport interne N-371. Centre de Géostatistique, École des Mines de Paris, Fontainebleau.

Jowett, G.H., 1955a. The comparison of means of industrial time series. *Applied Statistics* 4(1): 32–46.

Jowett, G.H., 1955b. The comparison of means of sets of observations from sections of independent stochastic series. *Journal of the Royal Statistical Society, Series B* 17(2): 208–227.

Kleingeld, W.J., Lantuéjoul, C., Prins, C.F., 2005. Sampling challenges in highly dispersed types of mineralisation. In: Holmes, R. (ed.) *Proceedings of the 2nd World Conference on Sampling and Blending*. Australasian Institute of Mining and Metallurgy, Brisbane, pp. 185–191.

Kolmogorov, A.N., 1941. The local structure of turbulence in incompressible viscous fluid at very large Reynolds' numbers. *Doklady Akademii Nauk SSSR* 30(4): 301–305.

Kracht, W., Emery, X., Paredes, C., 2013. A stochastic approach for measuring bubble size distribution via image analysis. *International Journal of Mineral Processing* 121: 6–11.

Krige, D.G., 1951. A statistical approach to some basic mine valuation problems on the Witwatersrand. *Journal of the Chemical, Metallurgical and Mining Society of South Africa* 52(6): 119–139.

Krige, D.G., 1952. A statistical analysis of some of the borehole values in the Orange Free State goldfield. *Journal of the Chemical, Metallurgical and Mining Society of South Africa* 53(3): 47–64.

Krige, D.G., 1997. A practical analysis of the effects of spatial structure and of data available and accessed, on conditional biases in ordinary kriging. In: Baafi, E.Y., Schofield, N.A. (eds.) *Geostatistics Wollongong' 96*. Kluwer Academic Publishers, Dordrecht, pp. 799–810.

Krige, D.G., Dunn, P.G., 1995. Some practical aspects of ore reserve estimation at Chuquicamata Copper Mine, Chile. In: McKee, D. (ed.) *Proceedings of the 25th International Symposium on Applications of Computers and Operations Research in the Mineral Industry*. Australasian Institute of Mining and Metallurgy, Brisbane, pp. 125–134.

La Pointe, P.R., Hudson, J.A., 1985. *Characterization and Interpretation of Rock Mass Joint Patterns*. Special Papers. Geological Society of America, Boulder, CO.

Langsæter, A., 1926. Om beregning av middelfeilen ved regelmessige linjetakseringer. *Meddelelser fra Det norske Skogforsøksvesen* 2(7): 5–47.

Lantuéjoul, C., 1990. *Cours de Sélectivité*. Cours C-140. Centre de Géostatistique, École des Mines de Paris, Fontainebleau, 72 p.

Lantuéjoul, C., 1991. Ergodicity and integral range. *Journal of Microscopy* 161(3): 387–403.

Lantuéjoul, C., 1997. Conditional simulation of object-based models. In: Jeulin, D. (ed.) *Advances in Theory and Applications of Random Sets*. World Scientific, Singapore, pp. 271–288.

Lantuéjoul, C., 2002. *Geostatistical Simulation: Models and Algorithms*. Springer, Berlin.

Lantuéjoul, C., Beucher, H., Chilès, J.P., Lajaunie, C., Wackernagel, H., 2005 Estimating the trace length distribution of fractures from line sampling data. In: Leuangthong, O., Deutsch, C.V. (eds.) *Geostatistics Banff 2004*. Springer, Dordrecht, pp. 165–174.

Lantuéjoul, C., Desassis, N., 2012. Simulation of a Gaussian random vector: a propagative version of the Gibbs sampler. Presented at: *Ninth International Geostatistics Congress, Oslo*, June 2012. Available at http://geostats2012.nr.no/pdfs/1747181.pdf (hal-00709250).

Lantuéjoul, C., Millad, M., 2008. Modelling the stone size distribution of a diamond deposit. In: Ortiz, J.M., Emery, X. (eds.) *Proceedings of the 8th International Geostatistics Congress*. Gecamin Ltda, Santiago, pp. 779–788.

Le Loc'h, G., 1990. *Déconvolution d'images Scanner*. Note N-8/90/G. Centre de Géostatistique, École des Mines de Paris, Fontainebleau.

Lerchs, H., Grossmann, I., 1965. Optimum design of open pit mines. *CIM Bulletin* 58(1): 47–54.

Li, W., 2006. Transiogram: a spatial relationship measure for categorical data. *International Journal of Geographical Information Science* 20(6): 693–699.

Li, W., 2007. Transiograms for characterizing spatial variability of soil classes. *Soil Science Society of America Journal* 71(3): 881–893.

Lindgren, F., Rue, H., Lindström, J., 2011. An explicit link between Gaussian fields and Gaussian Markov random fields: the stochastic partial differential approach. *Journal of the Royal Statistical Society B* 73(4): 423–498.

Linton, P., Browning, D., Pendock, N., Harris, P., Donze, M., Mxinwa, T., Mushiana, K., 2018. Hyperspectral data applied to geometallurgy. In: Becker, M. (ed.) *Geometallurgy Conference 2018 – Back to the Future*. Southern African Institute of Mining and Metallurgy, Cape Town, pp. 109–120.

Lotter, N.O., Kormos, L.J., Oliveira, J., Fragomeni, D., Whiteman, E., 2011. Modern process mineralogy: two case studies. *Minerals Engineering* 24(7): 638–650.

Madani, N., Emery, X., 2015. Simulation of geo-domains accounting for chronology and contact relationships: application to the Río Blanco copper deposit. *Stochastic Environmental Research and Risk Assessment* 29(8): 2173–2191.

Madani, N., Emery, X., 2017. Plurigaussian modeling of geological domains based on the truncation of non-stationary Gaussian random fields. *Stochastic Environmental Research and Risk Assessment* 31(4): 893–913.

Madani, N., Emery, X., 2019. A comparison of search strategies to design the cokriging neighborhood for predicting coregionalized variables. *Stochastic Environmental Research and Risk Assessment* 33(1): 183–199.

Maksaev, V., Munizaga, F., McWilliams, M., Fanning, M., Mathur, R., Ruiz, J., Zentilli, M., 2004. New chronology for El Teniente, Chilean Andes, from U/Pb, 40Ar/39Ar, Re-Os and fission track dating: implications for the evolution of a supergiant porphyry Cu-Mo deposit. In: Sillitoe, R.H., Perelló, J., Vidal, C.E. (eds.) *Andean Metallogeny: New Discoveries, Concepts and Updates*. SEG Special Publication 11. Society of Economic Geologists, Littleton, pp. 15–54.

Maleki, M., Emery, X., 2015. Joint simulation of grade and rock type in a stratabound copper deposit. *Mathematical Geosciences* 47: 471–495.

Maleki, M., Emery, X., 2017. Joint simulation of stationary grade and non-stationary rock type for quantifying geological uncertainty in a copper deposit. *Computers & Geosciences* 109: 258–267.

Maleki, M., Emery, X., Cáceres, A., Ribeiro, D., Cunha, E., 2016. Quantifying the uncertainty in the spatial layout of rock type domains in an iron ore deposit. *Computational Geosciences* 20(5): 1013–1028.

Maleki, M., Emery, X., Mery, N., 2017. Indicator variograms as an aid for geological interpretation and modeling of ore deposits. *Minerals* 7(12): 241.

Maleki, M., Jélvez, E., Emery, X., Morales, N., 2020. Stochastic open-pit mine production scheduling: a case study of an iron deposit. *Minerals* 10(7): 585.

Marcotte, D., Caron, J., 2013. Ultimate open pit stochastic optimization. *Computers & Geosciences* 51: 238–246.

Maréchal, A., 1982. Local recovery estimation for co-products by disjunctive kriging. In: Johnson, T.B., Barnes, R.J. (eds.) *Proceedings of the 17th International Symposium on Applications of Computers and Operations Research in the Mineral Industry* . Society of Mining Engineers of the AIME, Denver, CO, pp. 562–571.

Matérn, B., 1960. *Spatial Variation – Stochastic Models and Their Application to Some Problems in Forest Surveys and Other Sampling Investigations*. Meddelanden från Statens Skogsforskningsinstitut 49(5). Almaenna Foerlaget, Stockholm.

Matheron, G., 1962a. *Traité de Géostatistique Appliquée, Tome I*. Mémoires du Bureau de Recherches Géologiques et Minières, n°14. Éditions Technip, Paris.

Matheron, G., 1962b. *Traité de Géostatistique Appliquée, Tome III: les Phénomènes Transitifs*. Bureau de Recherches Géologiques et Minières, unpublished.

Matheron, G., 1963a. *Traité de Géostatistique Appliquée, Tome II: le Krigeage*. Mémoires du Bureau de Recherches Géologiques et Minières, n°24. Éditions BRGM, Paris.

Matheron, G., 1963b. Principles of geostatistics. *Economic Geology* 58(8): 1246–1266.

Matheron, G., 1965a. *Les Variables Régionalisées et leur Estimation: Une application de la Théorie des Fonctions Aléatoires aux Sciences de la Nature*. Masson, Paris.

Matheron, G., 1965b. *Remarque sur l'ouvrage de P. Gy « L'échantillonnage des Minerais en Vrac »*. Note géostatistique N°64. Centre de Géostatistique, École des Mines de Paris, Fontainebleau.

Matheron, G., 1967. *Le krigeage transitif*. Note géostatistique N°71. Centre de Géostatistique, École des Mines de Paris, Fontainebleau.

Matheron, G., 1971. *The Theory of Regionalized Variables and Its Applications*. Fasc. 5. Centre de Géostatistique, École des Mines de Paris, Fontainebleau.

Matheron, G., 1973. The intrinsic random functions and their applications. *Advances in Applied Probability* 5(3): 439–468.

Matheron, G., 1974. *Les fonctions de transfert des petits panneaux*. Note géostatistique n°127. Centre de Géostatistique, École des Mines de Paris, Fontainebleau.

Matheron, G., 1975a. Paramétrage de contours optimaux. Rapport N-403. Centre de Géostatistique, École des Mines de Paris, Fontainebleau.

Matheron, G., 1975b. Compléments sur le paramétrage des contours optimaux. Rapport N-401. Centre de Géostatistique, École des Mines de Paris, Fontainebleau.

Matheron, G., 1975c. Le paramétrage technique des réserves. Rapport N-453. Centre de Géostatistique, École des Mines de Paris, Fontainebleau.

Matheron, G., 1976. Forecasting block grade distributions: the transfer functions. In: Guarascio, M., David, M., Huijbregts, C. (eds.) *Advanced Geostatistics in the Mining Industry*. Kluwer Academic Publishers, Dordrecht, pp. 237–251.

Matheron, G., 1978. *Estimer et Choisir: Essai sur la Pratique des Probabilités*. Les Cahiers du Centre de Morphologie Mathématique de Fontainebleau, École des Mines de Paris, Fontainebleau.

Matheron, G., 1982. *Pour une analyse krigeante des données régionalisées*. Note N-732. Centre de Géostatistique, École des Mines de Paris, Fontainebleau.

Matheron, G., 1984. The selectivity of the distributions and the 'second principle of geostatistics'. In: Verly, G., David, M., Journel, A.G., Maréchal, A. (eds.) *Geostatistics for Natural Resources Characterization*. Reidel, Dordrecht, pp. 421–433.

Matheron, G., 1989. *Estimating and Choosing: An Essay on Probability in Practice*, 2nd edition by Presses des Mines, 2013. Springer-Verlag, Berlin.

Matheron, G., Beucher, H., de Fouquet, C., Galli, A., Guerillot, D., Ravenne, C., 1987. Conditional simulation of the geometry of fluviodeltaic reservoirs. SPE paper 16753 presented at the 62nd Annual Technical Conference and Exhibition of the Society of Petroleum Engineers. Society of Petroleum Engineers, Dallas, pp. 591–599.

Matheron, G., Serra, J., 2002. The birth of mathematical morphology. In: Talbot, H., Beare, R. (eds.) *Mathematical Morphology, Proceedings of the VIth International Symposium ISMM 2002*. CSIRO Publishing, Collingwood, pp. 1–16.

Melkumyan, A., 2015. Geometric modeling of the El Teniente Pipe: a new alternative. CFSG thesis. École des Mines de Paris, Fontainebleau.

Melkumyan, A., Nettleton, E., 2009. An observation angle dependent nonstationary covariance function for Gaussian process regression. In: Leung, C.S., Lee, M., Chan, J.H. (eds.) *Neural Information Processing. International Conference on Neural Information Processing 2009*. Lecture Notes in Computer Science, vol. 5863. Springer, Berlin, pp. 331–339.

Mery, N., Emery, X., Cáceres, A., Ribeiro, D., Cunha, E., 2017. Geostatistical modeling of the geological uncertainty in an iron ore deposit. *Ore Geology Reviews* 88: 336–351.

Monin, A.S., Yaglom, A.M., 1965. *Statisticheskaya gidromekhanika – Mekhanika Turbulenosti*. Nauka Press, Moscow.

Montoya, C., Emery, X., Rubio, E., Wiertz, J., 2012. Multivariate resources modelling for assessing uncertainty in mine design and mine planning. *Journal of the Southern African Institute of Mining and Metallurgy* 112: 353–363.

Moreno, E., Emery, X., Goycoolea, M., Morales, N., Nelis, G., 2017. Two-stage stochastic model for open pit mine planning under geological uncertainty. In: Dagdelen, K., (ed.) *Proceedings of the 38th Symposium of Applications of Computers and Operations Research in the Mineral Industry*. Colorado School of Mines, Golden, CO, pp. 13.27–13.33.

Müller, G., de Nordenflycht, R., 2006. Categorización de recursos minerales y reservas mineras – Codelco – Chile. In: Ortiz, J.M., Guzmán, R., Rubio, E., Henríquez, F., Lillo, P. (eds.)

Proceedings of the 2nd International Conference on Mining Innovation. Gecamin Ltda, Santiago, pp. 425–439.

Navarra, A., Menzies, A., Jordens, A., Waters, K., 2017. Strategic evaluation of concentrator operational modes under geological uncertainty. *International Journal of Mineral Processing* 164: 45–55.

Obukhov, A.M., 1941. O raspredelenii energii v spektre turbulentnogo potoka (On the distribution of energy in the spectrum of a turbulent flow). *Izvestiya Akademii Nauk SSSR, Seriya Geograficheskaya i Geofizicheskaya* 4–5: 454–456.

Obukhov, A.M., 1949a. Struktura temperaturnogo polya v turbulentnom potoke (Structure of the temperature field in a turbulent flow). *Izvestiya Akademii Nauk SSSR, Seriya Geograficheskaya i Geofizicheskaya* 13(1): 58–69.

Obukhov, A.M., 1949b. Lokalnaya struktura atmosphernoy turbulentnosty (Local structure of atmospheric turbulence). *Doklady Akademii Nauk SSSR* 67(4): 643–646.

Porcu, E., Bevilacqua, M., Genton, M.G., 2016. Spatio-temporal covariance and cross-covariance functions of the great circle distance on a sphere. *Journal of the American Statistical Association* 111(514): 888–898.

Priest, S.D., Hudson, J.A., 1976. Discontinuity spacings in rock. *Journal of Rock Mechanics and Mining Sciences & Geomechanics Abstracts* 13(5): 135–148

Priest, S.D., Hudson, J.A., 1981. Estimation of discontinuity spacing and trace length using scanline survey. *International Journal of Rock Mechanics and Mining Sciences & Geomechanics Abstracts* 18(3): 183–197.

Renard, D., Chilès, J.P., Rivoirard, J., Alfaro, M., 2013. Assessment of the resources of a gold deposit by transitive kriging. In: Costa, J.F., Koppe, J., Peroni, R. (eds.) *Proceedings of the 36th International Symposium of Applications of Computers and Operations Research in the Mineral Industry*. Fundação Luiz Englert, Porto Alegre, pp. 156–166.

Reyes, M., 2017. Operative mine planning, design and geological modeling: integration based on topological representations. PhD thesis in mining engineering, University of Chile.

Reyes, M., Morales, N., Emery, X., 2012. Final pit: simulated annealing approach with floating cones. In: Kuyvenhoven, R., Morales, J.E., Vega, C. (eds.) *Proceedings of the 5th International Conference on Innovation in Mine Operations – Book of Abstracts*. Gecamin Ltda, Santiago, p. 38.

Riquelme, A.J., Abellán, A., Tomás, R., 2015. Discontinuity spacing analysis in rock masses using 3D point clouds. *Engineering Geology* 195: 185–195.

Riquelme, R., Le Loc'h, G., Carrasco, P., 2008. Truncated Gaussian & plurigaussian simulations of lithological units in Mansa Mina deposit. In: Ortiz, J.M., Emery, X. (eds.) *Proceedings of the 8th International Geostatistics Congress*. Gecamin Ltda, Santiago, pp. 819–828.

Rivoirard, J., 1987. Two key parameters when choosing the kriging neighborhood. *Mathematical Geology* 19(8): 851–856.

Rivoirard, J., 1989. Models with orthogonal indicator residuals. In: Armstrong, M. (ed.) *Geostatistics*. Kluwer Academic Publishers, Dordrecht, pp. 91–107.

Rivoirard, J., 1994. *Introduction to Disjunctive Kriging and Non-linear Geostatistics*. Oxford University Press, Oxford.

Rivoirard, J., 2001. Weighted variograms. In: Kleingeld, W.J., Krige, D.G. (eds.) *Proceedings of the 6th International Geostatistics Congress*. Geostatistical Association of Southern Africa, Cape Town, pp. 145–155.

Rivoirard, J., 2004. On some simplifications of cokriging neighbourhood. *Mathematical Geology* 36(8): 899–915.

Rivoirard, J., Demange, C., Freulon, X., Lécureuil, A., Bellot, N., 2013. A top-cut model for deposits with heavy-tailed grade distribution. *Mathematical Geosciences* 45(8): 967–982.

Rivoirard, J., Freulon, X., Demange, C., Lécureuil, A., 2014. Kriging, indicators, and nonlinear geostatistics. *Journal of the Southern African Institute of Mining and Metallurgy* 114: 245–250.

Rivoirard, J. , Romary, T., 2011. Continuity for kriging with moving neighborhood. *Mathematical Geosciences* 43(4): 469–481.

Rivoirard, J., Simmonds, J., Foote, K.G., Fernandes, P., Bez, N., 2000. *Geostatistics for Estimating Fish Abundance*. Blackwell Science, Oxford.

Rondon, O., 2009. A look at plurigaussian simulation for a nickel laterite deposit. In: Dominy, S. (ed.) *Proceedings of the 7th International Mining & Geology Conference*. The Australasian Institute of Mining and Metallurgy, Melbourne, pp. 17–19.

Rossi, M.E., Parker, H.M., 1994. Estimating recoverable reserves: is it hopeless? In: Dimitrakopoulos, R. (ed.) *Geostatistics for the Next Century*. Kluwer Academic Publishers, Dordrecht, pp. 259–276.

Sánchez, L.K., Emery, X., Séguret, S.A., 2019. 5D geostatistics for directional variables: application in geotechnics to the simulation of the linear discontinuity frequency. *Computers & Geosciences* 133: 104325.

Sánchez, L.K., Emery, X., Widzyk-Capehart, E., 2018. Challenges in the geostatistical modelling of geotechnical and hydrogeological variables. In: Charrier, R., Valenzuela, M. (eds.) *Actas del XV Congreso Geológico Chileno: Geociencias Hacia la Comunidad*. Universidad de Concepción, Concepción, p. 610.

Schoenberg, I.J., 1942. Positive definite functions on spheres. *Duke Mathematics Journal* 9(1): 96–108.

Schwarzacher, W., 1969. The use of Markov chains in the study of sedimentary cycles. *Mathematical Geology* 1(1):17–39.

Séguret, S.A., 1988a. *Déconvolution d'images scanner – À la recherche des variogrammes ponctuels*. Note interne N-50/88/G. Centre de Géostatistique, École des Mines de Paris, Fontainebleau, 85 p.

Séguret, S.A., 1988b. Pour une méthodologie de déconvolution de variogrammes. Note interne N-51/88/G. Centre de Géostatistique, École des Mines de Paris, Fontainebleau, 36 p.

Séguret, S.A., 2011a. Block model in a multi facies context – application to a porphyry copper deposit. In: Beniscelli, J. (ed.) *Proceedings of the Second International Seminar on Geology for the Mining Industry*. Gecamin Ltda, Santiago, p. 10.

Séguret, S.A., 2011b. Spatial sampling effect of laboratory practices in a porphyry copper deposit. In: Alfaro, M., Magri, E., Pitard, F. (eds.) *Proceedings of the Fifth World Conference on Sampling and Blending*. Gecamin Ltda, Santiago, pp. 215–223.

Séguret, S.A., 2013. Analysis and estimation of multi-unit deposits: application to a porphyry copper deposit. *Mathematical Geosciences* 45(8): 927–947.

Séguret, S.A., 2015. Geostatistical comparison between blast and drill-holes in a porphyry copper deposit. In: Esbensen, K.H., Wagner, C. (eds.) *Proceedings of the 7th World Conference on Sampling and Blending*. IM Publishers, Chichester, pp. 187–192.

Séguret, S.A., 2016. Fracturing, crushing and directional concentration. *Mathematical Geosciences* 48(6): 663–685.

Séguret, S.A., Beniscelli, J., Carrasco, P., 2012. Cokriging partial grades – application to block modelling of copper deposits. Presented at the 9th International Geostatistics Congress, Oslo, Norway.

Séguret, S.A., Beucher, H., Chilès, J.P., Gesret, A., Noble, M., 2009. Geostatistical consulting, tutoring, development and implementation contract. Global Report of Year 4. Contrat ARMINES/CODELCO Chili. Mines ParisTech. R091212SSEG, 204 p.

Séguret, S.A., Beucher, H., Desassis, N., Fouedjio, F., Rivoirard, J., 2016. Geostatistical consulting, tutoring, development and implementation contract report of year 11 (2016). Contrat ARMINES/CODELCO Chili. Mines ParisTech. R161215SSEG, 227 p.

Séguret, S.A., Celhay, F., 2013. Geometric modeling of a breccia pipe – comparing five approaches. In: Costa, J.F., Koppe, J., Peroni, R. (eds.) *Proceedings of the 36th International Symposium on Application of Computers and Operations Research in the Mineral Industry*, Fundação Luiz Englert, Porto Alegre, pp. 257–266.

Séguret, S.A., Emery, X., 2019. *Géostatistique de Gisements de Cuivre Chiliens – 35 Années de Recherche Appliquée*. Presses des Mines, Paris.

Séguret, S.A., Goblet, P., 2016. Lithium deposit in deep aquifer: from resource evaluation to risk analysis. *V International Seminar Lithium in South America, Jujuy, Argentina, Panorama Minero* 438: 44–49.

Séguret, S.A., Goblet, P., Cordier, E., Galli, A., 2017. Bailer uncertainty evaluation in a lithium salar deposit. In: Dominy, S.C., Esbensen, K.H. (eds.) *Proceedings of the 8th World Conference on Sampling and Blending*. Australasian Institute of Mining and Metallurgy, Melbourne, pp. 395–402.

Séguret, S.A., Guajardo, C., 2015. Geostatistical evaluation of rock quality designation & its link with linear fracture. In: Schaeben, H., Tolosana Delgado, R., van den Boogaart, K.G., van den Boogaart, R. (eds.) *Proceedings of IAMG 2015 – 17th Annual Conference of the International Association for Mathematical Geosciences*. Curran Associates, Red Hook, NY, pp. 1043–1051.

Séguret, S.A., Guajardo, C., Freire, R., 2014. Geostatistical evaluation of fracture frequency and crushing. In: Castro, R. (ed.) *Proceedings of the 3rd International Symposium on Block and Sublevel Caving*. University of Chile, Santiago, pp. 280–288.

Serra, J., 1967. Échantillonnage et Estimation Locale des Phénomènes de Transition Miniers. Thèse de docteur-ingénieur, Tomes I & II, IRSID, Metz, Faculté des Sciences de Nancy.

Serrano, L., Vargas, R., Stambuk, V., Aguilar, C., Galeb, M., Holmgren, C., Contreras, A., Godoy, S., Vela, I., Skewes, M.A., Stern, C.R., 1996. The late Miocene to early Pliocene Río Blanco-Los Bronces copper deposit, Central Chilean Andes. In: Camus, F., Sillitoe, R.H., Petersen, R. (eds.) *Andean Copper Deposits: New Discoveries, Mineralization, Styles and Metallogeny*. Special Publication No. 5. Society of Economic Geologists, Littleton, CO, pp. 119–130.

Sillitoe, R.H., 1998. Major regional factors favoring large size, high hypogene grade, elevated gold content and supergene oxidation and enrichment of porphyry copper deposits. In: Porter, T.M. (ed.) *Porphyry and Hydrothermal Copper and Gold Deposits: A Global Perspective*. Australian Mineral Foundation, Glenside, pp. 21–34.

Skewes, M.A., Arevalo, A., Floody, R., Zuñiga, P., Stern, C.R., 2002. The giant El Teniente breccia deposit: hypogene copper distribution and emplacement. In: Goldfarb, R.J., Nielsen, R.L. (eds.) *Integrated Methods for Discovery: Global Exploration in the Twenty-First Century*. Society of Economic Geologists, Littleton, pp. 299–332.

Skewes, M.A., Arevalo, A., Floody, R., Zuñiga, P., Stern, C.R., 2006. The El Teniente megabreccia deposit, the world's largest deposit. In: Porter, T.M. (ed.) *Super Porphyry Copper and Gold Deposits – A Global Perspective*. Porter Geoscience Consultancy Publishing, Adelaide, pp. 83–113.

Skewes, M.A., Holmgren, C., Stern, C.R., 2003. The Donoso copper-rich, tourmaline-bearing breccia pipe in central Chile: petrologic, fluid inclusion and stable isotope evidence for an origin from magmatic fluids. *Mineralium Deposita* 38(1): 2–21.

Skvortsova, T., Armstrong, M., Beucher, H., Forkes, J., Thwaites, A., Turner, R., 2001. Applying plurigaussian simulations to a granite-hosted orebody. In: Kleingeld, W.J., Krige, D.G. (eds.) *Proceedings of the 6th International Geostatistics Congress (Geostats 2000 Cape Town)*. Geostatistical Association of Southern Africa, Cape Town, pp. 904–911.

Skvortsova, T., Beucher, H., Armstrong, M., Forkes, J., Thwaites, A., Turner, R., 2002. Simulating the geometry of a granite-hosted uranium orebody. In: Armstrong, M., Bettini, C., Champigny, N., Galli, A., Remacre, A. (eds.) *Geostatistics Rio 2000*. Kluwer Academic Publishers, Dordrecht, pp. 85–100.

Stagg, K.G., Zienkiewicz, O.C., 1968. *Rock Mechanics in Engineering Practice*. Wiley, New York.

Sturzenegger, M., Stead, D., Elmo, D., 2011. Terrestrial remote-sensing-based estimation of mean trace length, trace intensity and block size/shape. *Engineering Geology* 119(3–4): 96–111.

Subramanyam, A., Pandalai, H.S., 2008. Data configuration and the cokriging system: simplification by screen effects. *Mathematical Geology* 40(4): 425–443.

Talebi, H., Asghari, O., Emery, X., 2014. Simulation of the late-injected dykes in an Iranian porphyry copper deposit using the plurigaussian model. *Arabian Journal of Geosciences* 7(7): 2771–2780.

Talebi, H., Hosseinzadeh, E., MehdiAzadi, S., Emery, X., 2016. Risk quantification with combined use of lithological and grade simulations: application to a porphyry copper deposit. *Ore Geology Reviews* 75: 42–51.

Terzaghi, R.D., 1965. Sources of error in joint surveys. *Geotechnique* 5(3): 287–304.

Tuckey, Z., Stead, D., 2016. Improvements to field and remote sensing methods for mapping discontinuity persistence and intact rock bridges in rock slopes. *Engineering Geology* 208: 136–153.

Underwood, R., Tolwinski, B., 1998. A mathematical programming viewpoint for solving the ultimate pit problem. *European Journal of Operations Research* 107(1): 96–107.

van den Boogaart, K.G., Tolosana-Delgado, R., 2018. Predictive geometallurgy: an interdisciplinary key challenge for mathematical geosciences. In: Daya Sagar, B., Cheng, Q., Agterberg, F. (eds.) *Handbook of Mathematical Geosciences*. Springer, Cham, pp. 673–686.

Vann, J., Jackson, S., Bertoli, O., 2003. Quantitative kriging neighbourhood analysis for the mining geologist – a description of the method with worked case examples. In: Dominy, S. (ed.) *5th International Mining Geology Conference*. The Australasian Institute of Mining and Metallurgy, Melbourne, pp. 215–223.

Vargas, R., Gustafson, L.B., Vukasovic, M., Tidy, E., Skewes, M.A., 1999. Ore breccias in the Río Blanco-Los Bronces copper deposit, Chile. In: Skinner, B.J. (ed.) *Geology and Ore Deposits of the Central Andes*. Special Publication No. 7. Society of Economic Geologists, Littleton, CO, pp. 281–297.

Vargas, E., Morales, N., Emery, X., 2014. Footprint and economic envelope calculation for block/panel caving mines under geological uncertainty. In: Castro, R. (ed.) *Proceedings of the Third International Symposium on Block and Sublevel Caving*. University of Chile, Santiago, pp. 449–456.

Vergara, D., Emery, X., 2013. Conditional bias for multivariate resources estimation. In: Ambrus, J., Beniscelli, J., Brunner, F., Cabello, J., Ibarra, F. (eds.) *3rd International Seminar on Geology for the Mining Industry*. Gecamin Ltda, Santiago, pp. 27–33.

Verly, G., 1983. The multigaussian approach and its applications to the estimation of local reserves. *Mathematical Geology* 15(2): 259–286.

Wackernagel, H., 2003. *Multivariate Geostatistics: An Introduction with Applications*. Springer-Verlag, Berlin.

Warburton, P.M., 1980. A stereological interpretation of joint trace data. *International Journal of Rock Mechanics and Mining Sciences & Geomechanics Abstracts* 17(4): 181–190.

Weissmann, G.S., Fogg, G.E., 1999. Multi-scale alluvial fan heterogeneity modeled with transition probability geostatistics in a sequence stratigraphic framework. *Journal of Hydrology* 226(1–2): 48–65.

Wills, B.A., Finch, J., 2015. *Wills' Mineral Processing Technology: An Introduction to the Practical Aspects of Ore Treatment and Mineral Recovery*, 8th edition. Elsevier, Amsterdam.

Wright, E., 1989. Dynamic programming in open pit mining sequence planning: a case study. In: Weiss, A. (ed.) *Proceedings of the 21st International Symposium of Applications of Computers and Operations Research in the Mineral Industry*. Society of Mining Engineers, Littleton, CO, pp. 415–422.

Yaglom, A.M., 1957. Some classes of random fields in n-dimensional space, related to stationary random processes. *Theory of Probability and Its Applications* 2(3): 273–320.

Yaglom, A.M., 1958. Correlation theory of processes with stationary random increments of order n. *American Mathematical Society Translations* 2: 8–87.

Yaglom, A.M., Pinsker, M.S., 1953. Random processes with stationary increments of order n. *Doklady Akademii Nauk SSSR* 90(5): 731–734.

Yunsel, T.Y., Ersoy, A., 2011. Geological modeling of gold deposit based on grade domaining using plurigaussian simulation technique. *Natural Resources Research* 20(4):1–19.

Yunsel, T.Y., Ersoy, A., 2013. Geological modeling of rock type domains in the Balya (Turkey) lead-zinc deposit using plurigaussian simulation. *Central European Journal of Geosciences* 5(1): 78–89.

Zhang, L., Einstein, H.H., 2000. Estimating the intensity of rock discontinuities. *International Journal of Rock Mechanics and Mining Sciences* 37: 819–837.

Index

Printed in the United States
By Bookmasters